LAW AND CATASTROPHE

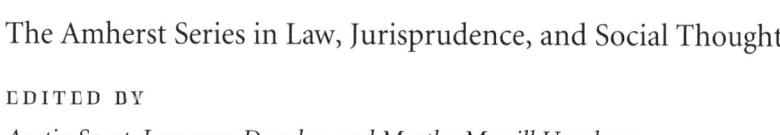

The Amherst Series in Law, Jurisprudence, and Social Thought

EDITED BY

Austin Sarat, Lawrence Douglas, and Martha Merrill Umphrey

Law and Catastrophe

Edited by

AUSTIN SARAT

LAWRENCE DOUGLAS

MARTHA MERRILL UMPHREY

STANFORD UNIVERSITY PRESS

Stanford, California, 2007

Stanford University Press
Stanford, California
© 2007 by the Board of Trustees of the
Leland Stanford Junior University
All rights reserved
Printed in the United States of America

Library of Congress Cataloging-in-Publication Data

Law and catastrophe / edited by Austin Sarat, Lawrence Douglas, Martha Merrill Umphrey.
    p.   cm.—(The Amherst series in law, jurisprudence, and social thought)
    Includes index.
    ISBN 978-0-8047-5683-9 (cloth : alk. paper)
    1. Disasters—Law and legislation.   2. Disaster relief—Law and legislation.   3. Industrial accidents—Law and legislation.   4. Social responsibility of business.   5. Disasters.   I. Sarat, Austin.   II. Douglas, Lawrence.   III. Umphrey, Martha Merrill.

K1980.L39 2007
344.05'34—dc22

                                                          2007005949

Typeset by Newgen–Austin in 10/14.5 Minion

To my son Ben (A.S.)

Acknowledgments

The essays in this book were originally prepared for and presented as a seminar series at Amherst College during the 2003–2004 academic year. We are grateful to our Amherst College colleagues David Delaney and Nasser Hussain for their enthusiastic and helpful participation in that series. We thank our students in Amherst College's Department of Law, Jurisprudence, and Social Thought for their interest in the issues addressed in this book. Finally, we would like to express our appreciation for generous financial support provided by the College's Charles Hamilton Houston Forum on Law and Social Change.

Contents

Contributors

LAWRENCE DOUGLAS is Professor of Law, Jurisprudence, and Social Thought at Amherst College.

LINDA ROSS MEYER is Professor of Law at Quinnipiac Law School.

RAVIT PE'ER-LAMO REICHMAN is Assistant Professor of English at Brown University.

AUSTIN SARAT is the William Nelson Cromwell Professor of Jurisprudence and Political Science and Professor of Law, Jurisprudence, and Social Thought at Amherst College and Five College Fortieth Anniversary Professor.

SYLVIA SCHAFER is Professor of History at the University of Connecticut.

RONEN SHAMIR is Professor of Sociology at Tel-Aviv University.

MARTHA MERRILL UMPHREY is Associate Professor of Law, Jurisprudence, and Social Thought at Amherst College.

JAMES E. YOUNG is Professor of English as well as Judaic and Near Eastern Studies at the University of Massachusetts.

LAW AND CATASTROPHE

A Jurisprudence of Catastrophe:
An Introduction

LAWRENCE DOUGLAS

AUSTIN SARAT

MARTHA MERRILL UMPHREY

The Idea of Catastrophe

The study of catastrophe is a growth industry. Today, cosmologists scan the heavens for asteroids of the kind that smashed into Earth some 90 million years ago, leading to the swift global die-off of the dinosaurs. Climatologists create elaborate models of the chaotic weather and vast flooding that will result from the continued buildup of greenhouse gases in the planet's atmosphere.[1] Epidemiologists plan for the next pandemic of the proportions of the Spanish flu, which traversed the world in the waning days of the Great War and left a trail of 20 million dead.[2] Physicists ponder the chances that experiments with subatomic particles might lead to a "strangelet event": the sudden collapse of the planet into a hyperdense sphere the diameter of a football field.[3] Meanwhile, terrorist experts and homeland security consultants struggle to prepare for a wide range of possible biological, chemical, and radiological attacks: aerated smallpox virus spread by a crop duster, botulism dumped into an urban reservoir, a dirty bomb detonated in a city center.[4]

These events share the quality of being merely *possible*, but recent headlines supply more than sufficient examples of real-world catastrophes: Think of the ongoing humanitarian tragedy unfolding in the Darfur region of Sudan, the mass death that resulted from the tsunami of Banda Aceh, or the devastation visited on New Orleans by Hurricane Katrina. But if catastrophes run from the actual and the inevitable to the speculative and the highly improbable, what features do these events share such that they all can be denoted by the same term? Our brief parade of examples suggests that the term does not strictly limit itself to either natural occurrences or events caused by humans; it encompasses both. Nor does the word suggest a specific temporal dimension: Catastrophes may be sudden, caused by the strike of a storm or a terrorist, or slowly unfolding, the result of global warming or a politics of racial exclusion culminating in genocide. And the term is more than simply a synonym for disaster: *Catastrophe* denotes something qualitatively more serious; we might agree that all catastrophes are disasters, but not all disasters are catastrophes. Catastrophe, then, is a limit term—it names a condition at the frontier and endpoint of all forms of fiasco and calamity.

In the original Greek, the term *catastrophe* denotes total "ruin" or a radical "overturning." Yet an overturning of what? As our examples suggest, we reserve catastrophe to describe events that bring about a substantial, if not mass, loss of life. Such events usually also include widespread destruction of infrastructure, collapse of public services, and massive disruptions of quotidian routine and existence. As radical disruptions of conventional processes and routines of life, catastrophes can be understood as overturning the *very concept of order itself*. Yet even this can be taken one step further, for by upsetting, if not utterly destroying, the predicates of ordered existence, catastrophe can also be understood as overturning the very belief in *normative order*—the idea that life should and can be patterned according to a system of rules.

It is this quality—the erosion of a belief in normative order—that most strikingly characterizes the response of writers and thinkers to earlier catastrophes. In the wake of the earthquake that devastated Lisbon on All Saints' Day in 1755, an event which by all accounts traumatized the Enlightenment imagination, Voltaire penned these famous lines:

> Will you say: "This is the result of eternal laws
> Directing the acts of a free and good God!" . . .
> Did Lisbon, which is no more, have more vices
> Than London and Paris immersed in their pleasures?
> Lisbon is destroyed, and they dance in Paris![5]

Move forward to the paradigmatic catastrophe of the twentieth century—the Nazi's destruction of the European Jews, denoted in Hebrew simply as the Shoah, "the Catastrophe"[6]—and we find a similar crisis of faith. In his haunting memoir *Night*, Elie Wiesel describes how Nazi policies destroyed both human beings and the possibility of a belief in God: "Never shall I forget those moments which murdered my God."[7] In *If This Is a Man*, Primo Levi echoes Wiesel's words in his remarkable description of the outrage he feels toward an inmate who, in earshot of those "selected" to be gassed, openly praises God for sparing his life: "If I was God, I would spit at Kuhn's prayer."[8]

Bringing the Law In

Framed in these terms, what would it mean to speak of a *jurisprudence* of catastrophe? If the question sounds unfamiliar, it is for the simple reason that little work has been done theorizing the relationship between law and catastrophe. The relationship between law and other limit conditions—such as states of emergency—has been the subject of a rich and growing literature.[9] By contrast, little has been written about law and catastrophe, and in devoting a volume to the subject, our hope is less to provide an overview of a well-defined field than to sketch the contours of a relatively fresh, yet crucial, terrain of inquiry.

As a preliminary matter, we might think of the relationship between law and catastrophe in two completely different ways. In the first, catastrophe could be considered as *issuing* from law. If modern secular thought sees catastrophe as an overturning of normative order, traditional religious thought often viewed catastrophe as an *expression of normativity*. In this understanding, catastrophe was seen as an instrument of law, as a tool of justice, a form of divine sanction for the violation of God's law.[10] In biblical stories such as Noah's Ark and the destruction of Sodom and Gomorrah, catastrophe is viewed in precisely this manner. In the *Prophets*, we read:

> For three crimes of Damascus, and for four,
> I have decided irrevocably!
> ... I will send a fire into the house of Hazael ...[11]

Far from a random event, catastrophe is God's tool for punishing the human community in a manner that is lawful, meaningful, and even just. Of course, this is the view attacked by Voltaire when he writes, *Will you say: "This is the result of eternal law / directing the acts of a free and good God!"* But notwithstanding the secular humanist critique, we continue to hear echoes of the divine understanding in pronouncements of contemporary religious fundamentalists of all stripes.[12]

At the other pole from the divine view is the liberal account. In this genealogy, law issues from catastrophe, not vice versa. In its most influential iteration, in Hobbes's *Leviathan*, the liberal account describes a state of catastrophic disorder from which all prudent reasoning persons seek to flee.[13] Here, however, we need to justify the designation of the state of nature as a state of catastrophe. By our own reckoning, catastrophe is an overturning of a preexisting order. The state of nature, by contrast, appears to preexist order, and thus it cannot be said to overturn anything. Yet this challenge overlooks a key aspect of Hobbes's argument. Recall that, for Hobbes, the state of nature does *not* refer to a historical condition; in this regard, he parts company with Locke, who believed that all societies evolved out of such a state. For Hobbes, the state of nature is an *analytic* condition: It is the state that societies always threaten to revert back to, given the right set of conditions.

Seen in this light, law is constituted in the effort to escape catastrophe; indeed, law is what makes possible the defeat of catastrophic disorder and violence. In this reckoning catastrophe is both juris generative[14]—it is the ever-present threat of chaos that creates the need for law—and the very antithesis or negation of law—it is the uncontrollable force that threatens to extirpate law's ordering effects on social life. Once law has been established to maintain social order, catastrophe remains law's nemesis, the unruly force that would overturn the rules and regimes so carefully constructed by the principles and practices of legality. In this picture, the specter of catastrophe plays a crucial role in law's justificatory logic, as law appears as the bulwark between civilization and catastrophic disorder.

Both of these understandings—the divine and the liberal—appear in Linda Ross Meyer's contribution to our collection, "Catastrophe: Plowing Up the Ground of Reason." Meyer explores how modern law strives to gain dominion over catastrophe; how it attempts to subdue, domesticate, and colonize it. Specifically, Meyer shows how law works to master catastrophe through its standard responses to disorder—through strategies of *anticipation, prevention*, and *amelioration*. Yet as Meyer points out in her close reading of the Book of Job, this was not always the case. Historically, catastrophe was conceptualized not simply as refractory to secular law's regulatory techniques but as definitionally beyond law's regulatory domain. Enfolded within the category of "acts of God"—disasters born of an opaque and inaccessible divine logic—catastrophes were viewed as "unforeseeable" events for which, say, an insurer could not be held legally liable. Rather, they were seen as beyond secular law's jurisdiction. As Meyer makes clear, over time this understanding changed. With the ascendancy of liberal jurisprudence, law came to expand its jurisdictional ambit: Catastrophe was no longer seen as beyond the law's power of control but as a challenge to it—a challenge that demanded a legal response.

According to Meyer, this jurisdictional shift left neither law nor catastrophe intact. In *The Faces of Injustice*, Judith Shklar observed that the distinction between misfortune and injustice is not natural, inevitable, or stable.[15] Shklar insisted that the traditional idea—that misfortunes refer to accidents beyond human calculation and control, whereas injustices refer to harms that result from negligent, reckless, or intentionally wrongful human actions—is no longer sustainable in a liberal legal community. In our legal universe, the failure to ameliorate the suffering of others may transform what first seemed a "misfortune" into an "injustice." Meyer's chapter powerfully echoes and expands upon Shklar's argument. Indeed, as several of our contributors make clear, one of the distinguishing features of catastrophe is its *urgent call for response*. Commonplace misfortunes can be ignored easily enough. The state can figuratively step over the body of the homeless without triggering a crisis of legitimacy. Catastrophe, by contrast, tolerates no such absence of response.

Within the liberal state, catastrophe demands action, and the failure to act will be seen as a crucial element of catastrophe itself. In the case of Hurricane Katrina, it was the government's failure to respond, more than the storm surge itself, that was decried as catastrophic.[16] The chaos, desperation, and lawlessness that engulfed New Orleans came to be seen as only *indirectly* a consequence of the hurricane and flood; they were the *direct* result of the government's failure to mobilize an effective response. Many critics of the government's response went further still, locating the crucial failure not in the woefully bungled relief efforts but in the failure to prevent the catastrophe from occurring in the first place. Had the government invested in the construction of a more durable levee system—and it was probable, if not certain, that New Orleans would one day be pummeled by a category-four hurricane[17]—then the flood damage might have been completely averted. Indeed, such arguments echo Rousseau's famous reaction to Voltaire's response to the Lisbon earthquake of 1755. Far from seeing the catastrophe as evidence of a disordered and meaningless universe, Rousseau understood the event in terms of human failure: ". . . it was hardly nature who assembled there twenty-thousand houses of six or seven stories. If the residents of this large city had been more evenly dispersed and less densely housed, the losses would have been fewer or perhaps none at all."[18]

In Rousseau's argument, we hear the position that Meyer describes as the colonizing impulse of the law—the desire to wrestle catastrophe from the realm of "acts of God" and to place it under the regulatory auspices of secular law. Thus, as we begin to theorize the relationship between law and catastrophe, Meyer's contribution frames two additional insights. First, *pace* Shklar, the distinction between catastrophes caused by nature and those orchestrated by humans is neither clear nor stable. Second, it is in the nature of catastrophe that any post hoc response will tend to be viewed as insufficient and belated. In terms of the triumvirate of ordering strategies available to the law, catastrophe demands that law be geared to anticipation and prevention rather than amelioration. When the law finds itself in the position of mobilizing ameliorative actions, it has already failed in important respects to master the challenge of catastrophe.

Anticipation and Prevention

Regulation can be understood as the "act or process of controlling by rule or restriction."[19] In our contemporary legal universe, the concept of regulatory law specifically refers to orders or rules enforced by administrative agencies. These agencies are often responsible for articulating and enforcing rules that attempt to anticipate and prevent catastrophes.[20] The Securities and Exchange Commission, for example, establishes, interprets, and enforces rules designed to prevent a catastrophic collapse of the stock market. The Transportation Security Administration is responsible for regulating passenger and freight transportation in the United States in a manner that will safeguard against catastrophic accidents and attacks. Although most regulatory agencies tend to focus on anticipation and prevention, this is not always the case. As an administration specifically geared to disaster relief and assistance, the Federal Emergency Management Agency stands as the prime example of a regulatory administration with a focus on amelioration.

Yet regulatory agencies and administrations alone do not occupy the field of law's response to catastrophe. Viewed conceptually, it is helpful to distinguish between regulatory strategies (and by this we do not mean to limit our discussion to the world of administrative law) that involve criminal law and those that rely on civil law. *Criminal* law's response has been preoccupied with two kinds of catastrophic threats: those posed by states and those posed by terrorist groups. Under the former, we think of trials of perpetrators of state-sponsored atrocities such as genocide and crimes against humanity. Yet, as we'll see when we examine the concluding contributions to our volume, it is often difficult to say whether such actions are, at their core, preventive or ameliorative. In the case of terrorist threats, the law's response has clearly been aimed at anticipation and prevention, though this itself has emerged as a subject of intense controversy. Indeed, the most pressing constitutional issues of the day now involve questions such as the following: Does the executive branch enjoy inherent powers to order wiretaps of alleged terrorist suspects in the absence of congressional authorization or judicial warrants? Does the executive have inherent powers to authorize the use of "unorthodox" interrogation techniques for terrorist suspects?[21] Should terrorist suspects be entitled to the full panoply of rights and procedures that come with trials before Article III courts?[22] And how can we characterize the powers claimed by the executive branch? Are they either legal or illegal, or is it important to use another characterization altogether such as "extralegal"?[23] These controversies raise the larger question of how we should go about striking the proper balance between civil liberties and collective security in an age of terrorism.

However we might think about these matters, we must recall that it is not terrorism per se that creates the need for a recalibration of what might be considered law's most critical balance. After all, anarchist groups were widespread at the turn of the twentieth century, and the terror they spawned was considered of epidemic levels.[24] New, then, is not the existence of terrorism but the *nature* of the threat that it poses. As Michael Ignatieff's *The Lesser Evil*[25] and Richard Posner's *Catastrophe: Risk and Response*[26] make clear, what distinguishes today's terrorists is their power to inflict catastrophic damage, which was the lesson learned on 9/11. It is this fact—the specter of catastrophe—that has changed the terms of the debate for those on the left and the right. At the most basic level, the need to anticipate and prevent terrorist-sponsored catastrophes has raised foundational questions about the substance and procedures of criminal law. Predictably enough, the answers to these questions often fall back on classic Hobbesian arguments: that in the face of catastrophic violence, the interests of security trump all. And yet this logic is peculiarly self-defeating: The law's draconian efforts to anticipate and prevent catastrophic terrorism threaten to erode law's distinctive status as a *normative* tool of social order.

As we shift our attention to civil law—from the topic of terror to the problem of error—the strategies for anticipating and preventing catastrophe are perhaps less controversial. In the world of civil law, catastrophic threats originate not in Al Qaeda plots but in corporate malfeasance. The paradigmatic instance remains the Bhopal disaster,[27] and the paradigmatic response involves the assignment of risk: The law shifts risk to the party best able to assess and manage it. This is the "costs of accidents" approach pioneered by Guido Calabresi and now familiar to all students of torts.[28] Calabresi's approach can be seen as a nice illustration of the historical development described by Meyer, as tort law comes to abandon the categorical and jurisdictional approach to calamity (acts of God) and instead enshrines the principle of *foreseeability*. Yet as Meyer implicitly suggests, the cost of accidents applies only imperfectly in the case of catastrophes. Calabresi's system, we recall, remains largely agnostic with respect to the question of prevention or amelioration.

From the standpoint of strict social efficiency, the decision to invest, say, in a safer workplace enjoys no particular advantage over the decision to pay damages for workplace-related injuries. In the case of catastrophe, however, the costs of amelioration are so staggeringly high that emphasis must be on prevention.

This also is a central argument of Posner's *Catastrophe: Risk and Response*. For Posner, the law often falters in response to catastrophe as a result of the "bafflement that most people feel when they try to think about events that have an extremely low probability of occurring even if they will inflict enormous harm if they do occur."[29] At times, Posner inadvertently underscores the difficulties, and the grotesqueries, that arise from the attempt to calculate catastrophic risks. Balancing the potential benefits of a series of planned physics experiments at a new laboratory at Brookhaven against their potential costs, Posner calculates as follows:

> RHIC's (Brookhaven's Relativistic Heavy Ion Collider) net present value would be $400 million. This figure is arrived at by subtracting from $2.1 billion—the present value of a stream of annual benefits of $250 million for 10 years discounted at 3 percent—$1.1 billion, the present value of the annual operating costs, similarly discounted, and $600 million, the accelerator's fixed costs.
>
> But now suppose the cost of extinction of the human race, which . . . can very conservatively be estimated at $600 trillion . . .[30]

Clearly, there is something farcical about this reasoning, though Posner delivers it with a straight face. Still, it underscores the centrality of the claim that the specter of catastrophe upsets the costs of accidents formulae; in responding to the challenges posed by catastrophe, civil law must emphasize prevention rather than amelioration.

At first blush, this point finds strong support in Ronen Shamir's contribution, "Catastrophes and Humanitarian Corporate Responsibility: A Conceptual Critique." Shamir explores the relationship between multinational corporations (MNCs) and humanitarian catastrophes, such as the unfolding genocide in the Darfur region of Sudan. Although many commentators have argued in support of submitting the atrocities of Darfur to a criminal tribunal,[31] Shamir is less interested in the perpetrators than in their corporate allies. To his credit, Shamir rejects the standard antiglobalist critique that lays the blame for all the world's disasters at the doorstop of the MNCs. At the same time, he is highly critical of the prevailing neoliberal ideology, which views the exercise of corporate social responsibility—the decision of a parent company to divest itself of a subsidiary doing business with a criminal regime, for example—as largely a matter of civic virtue and voluntary action. Against this prevailing ideology, which also informs the larger logic of humanitarian relief, Shamir exhorts us to use the law to prod MNCs to do more to prevent and ameliorate catastrophe.

Yet beneath Shamir's normative argument, and almost concealed by it, lies a striking and troubling insight. Like Meyer, Shamir reminds us of the instability of the distinction between natural and human-made catastrophes; he notes, for example, that "it is now widely acknowledged that famines are not the result of natural causes alone" (p. 46). Instead, they typically result from a confluence of factors: political corruption, administrative ineptitude, failed infrastructure, deficient planning, and so forth. But from here, Shamir goes on to destabilize the distinction between prevention and amelioration—or to use the corollary terms in the world of MNCs, between *development* and *relief*—that informs Posner's jurisprudence of catastrophe.

Drawing on the vast literature of "dependent development,"[32] Shamir points out that efforts at development may result in propping up corrupt regimes that perpetrate grave humanitarian abuses. On the other hand, relief efforts in the absence of development (e.g., sending food to a region devastated by famine) may do little to help the people on the ground and may simply contribute to the vicious logic of socioeconomic dependence. Worse yet, efforts to address one catastrophe may only exacerbate another: Imagine, for example, a program designed to eliminate catastrophic levels of infant mortality that inadvertently contributes to catastrophic overpopulation: How do we assess the efficacy and justice of such initiatives? Posner's work on catastrophe assumed a relatively stable distinction between prevention and amelioration and offered seemingly sound reasons for devoting resources to the former rather than the latter. Shamir's chapter, however, problematizes these neat and well-meaning policy recommendations. Law's efforts to respond to the threat of calamity may lead it to precipitate the very thing it is trying to prevent.

This conundrum that law's efforts at prevention and amelioration may only exacerbate the problem is likewise the focus of Sylvia Schafer's chapter, "Political Catastrophe and Liberal Legal Desire: Two Stories of Revolution, Remediation, and Return from the French Nineteenth Century." While Meyer examined the ways in which law attempts to colonize catastrophe and render it judicially manageable, a mere blip in the grid of law's order, Schafer examines the ways in which catastrophe upends legal order in unanticipated ways. In her reading of Balzac's *Colonel Chabert*, she examines how the law's tools of colonization—actuarial thinking, policies of indemnification, principles of risk management—all backfire in spectacular fashion in the case of the novel's eponymous protagonist. Here, law's attempts to mitigate the devastation of the Napoleonic Wars end up destroying what they aim to repair. Unable to deal with the resurfacing of a lost soldier deemed legally dead, the law succeeds where war failed—at utterly destroying the survivor.

At the same time that Schafer's reading reminds us of the fragility of legal personhood, it also supports Shamir's implicit challenge to Posner. In the second half of her essay, Schafer turns her attention away from the dislocations of the Napoleonic Wars to the trauma of the revolutions of 1848. As violence gripped the continent, the law was faced with a dilemma: How was it to deal with spasms of lawless uprisings by the impoverished classes? Formally conceived, this problem perfectly recapitulates and anticipates Shamir's discussion of the social responsibility of multinational corporations. Threatened with the specter of large-scale chaos in 1848, some social reformers urged the path of "relief": They believed that by providing legal assistance to the poor, the impoverished class could find help in the courts and so would renounce violence in the streets. These reformers, however, were opposed by those who resisted liberalizing access to the law. Doing so, they feared, without fundamentally altering the underlying social conditions of poverty, would only embolden the poor to more violence, creating greater social upheaval. In Shamir's terms, "relief" without "development" would only lead to greater social instability. And so Schafer locates a crisis, if not a tragedy, at the heart of the liberal legal project. As the attempt to help the socially disadvantaged threatens to mobilize a "predatory collectivity," the law becomes paralyzed by the specter that its gestures of amelioration will only create greater chaos. Legal efforts to contain catastrophe are haunted by the fear that they will only aggravate it.

Amelioration

We have already observed that the nature of catastrophe suggests that law's energy be devoted to anticipation and prevention rather than amelioration. Yet we have also seen that the distinction between prevention and amelioration is neither neat nor stable. Further, we have noted that attempts to ameliorate catastrophe will always remain troubled by the specter of insufficiency and belatedness, if not plagued by the concern that ameliorative gestures often exacerbate harm. Does this mean that law has no ameliorative role to play in the wake of catastrophe? This is the question addressed in the chapters by our final two contributors, Ravit Pe'er-Lamo Reichman and James E. Young.

Both Reichman and Young examine the nature of the law's response to the Holocaust, perhaps the most extraordinary and emblematic catastrophe of the twentieth century. As we've seen, Meyer argues that catastrophes were once viewed as acts of God, beyond law's jurisdictional reach, but that modern law resists this approach, claiming dominion over catastrophe. Meyer's claim again finds provocative support in the case of international law's response to Nazi genocide. In an effort to bring Nazi perpetrators to justice, international jurists articulated novel incriminations, such as the crime against humanity and the crime of genocide. They also pioneered the use of unorthodox jurisdictional theories, such as universal jurisdiction.[33] Such innovations were specifically intended to bring catastrophic crimes within law's jurisdictional reach. The success of these gestures is borne out by the fact that when we think of the law's response to the Shoah, we can't help but conjure images of Goering on trial at Nuremberg or of Eichmann in the glass booth at Jerusalem.

The Holocaust thus presents a prime example of catastrophe prompting law to innovate. But how do we characterize the nature and purposes of these innovations, and how do we measure "success" in terms of the prevention–amelioration continuum that we encounter throughout our volume? This question loomed large at the Nuremberg Trial, the subject of Reichman's chapter. Before, during, and after the historic trial before the International Military Tribunal, commentators asked whether any legally imposed sanction could ever right the scales of justice against the terrible weight of crimes against humanity and genocide. In his famous opening statement before the International Military Tribunal, chief allied prosecutor Robert Jackson addressed this issue when he observed, "In the prisoners' dock sit twenty-odd broken men . . . It is hard now to perceive in these men as captives the power by which as Nazi leaders they once dominated much of the world and terrified most of it. Merely as individuals their fate is of little consequence to the world."[34] Hannah Arendt, years before she essayed her famous banality of evil thesis, also spoke to the adequacy of a legal response in a letter to Karl Jaspers: "It may be essential to hang Goering, but it is totally inadequate. That is, this guilt, in contrast to criminal guilt, oversteps and shatters any and all legal systems. That is why the Nazis in Nuremberg are so smug."[35]

Given the unbridgeable gap between crime and sanction, what ends are served by submitting catastrophic crimes to legal judgment? Certainly, there are those who attempt to justify such a response in terms of conventional ideas of prevention— that is, deterrence. The International Criminal Tribunal for the former Yugoslavia[36] and the International Criminal Tribunal for Rwanda[37] invoke this justification in their enabling language, and the statute of the International Criminal Court specifically speaks of the determination "to put an end to impunity for the perpetrators of these crimes and thus to contribute to the prevention of such crimes."[38] Many observers, however, remain unconvinced by the deterrence justification.[39] Nuremberg, for example, did little to deter Pol Pot or Slobodan Milosevic, and Milosevic's trial, in turn, obviously made little impression upon the rebel leaders in Sudan.

Other scholars, by contrast, defend these processes as tools of amelioration rather than of prevention. Such trials, it is argued, provide a needed account of the history of a catastrophic episode and offer a valuable forum in which memories of survivors and witnesses can be publicly shared, recognized, and honored.[40] By serving the interests of history and memory, these proceedings can help persons, groups, and nations come to terms with the dreadful legacy of catastrophe.

It is precisely Nuremberg's relationship to memory that is the focus of Reichman's essay, "Committed to Memory: Rebecca West's Nuremberg." Reichman comes to the trial through the reportage of Rebecca West, the famous British novelist and writer who covered the trial for London's *Daily Telegraph*. If, at the Eichmann trial, Arendt was drawn to the prosecution's failure to comprehend the character of the accused,[41] West was drawn to a very different feature of the Nuremberg Trial: its boredom. West suffers no writer's block in characterizing this boredom; she returns to it time and again, capturing the quality of dullness with the most vivid language and images: "the courtroom was a citadel of boredom"; "Nuremberg . . . was . . . water-torture, boredom falling drop by drop on the same spot of the soul"; "the symbol of Nuremberg was a yawn"; "this was boredom on a huge historic scale."[42]

Yet as Reichman makes clear, West doesn't locate Nuremberg's boredom in any particular legal failing (e.g., the ineptitude of the prosecutors or the long-windedness of the defense counsels). Rather, she locates the source of Nuremberg's boredom in *disappointed expectations*: the first international war crimes trial failed to deliver a spectacle commensurate with the crimes to be judged. Put another way, Nuremberg was boring because, alas, Nuremberg was a trial. This parsing invites us to see disappointed expectations as a structural feature of criminal law's contact with catastrophe: This is the corollary of our observation that all attempts at amelioration will remain belated, incomplete, and unsatisfactory. Of course, in certain respects, it is possible to see the dullness of Nuremberg as a legal success. If one of the purposes of trials of catastrophe is to reintroduce norms of legality into a radically lawless space, the very dryness of the proceeding can be construed as a triumph of legal sobriety over lawless chaos.

Yet in Reichman's reading, West is less concerned with interpreting the meaning of Nuremberg's boredom than in using the experience as a means of arriving at a fuller phenomenology of the trial. For West, the experience of boredom was deeply productive, as it created the space for memory to take root and grow. What kind of memory? Certainly not the kind of "responsible memory" that the Eichmann trial strove to encourage. By permitting survivors to testify in open court, the Eichmann trial aimed to make the trial's spectators into witnesses of the witnesses, now morally burdened with the memory of catastrophe narrated in a juridical setting.[43] For West, however, memory is not forged through the confrontation with acts of tragic witnessing. Instead, it is the experience of boredom that permits West to locate the voice of memory not simply in the courtroom and during the proceedings but just as crucially outside the courtroom, in encounters with citizens of Nuremberg. The vacancy of the words uttered in court thus makes her all the more attentive to the voices of gardeners, servants, former soldiers—Germans with whom she has chance, fleeting, yet resonant encounters during her perambulations in the ruined city. The experience of boredom invites a distinctly modernist form of reportage in which memory is forged out of the collision of multiple and conflicting voices. The memory of catastrophe cannot be controlled or dictated by the prosecution's script. Rather, it emerges out of the dialogue created between voices inside and outside the courtroom. Through these juxtapositions and shifts of perspective, we gain insight into the meaning of law's confrontation with catastrophe at Nuremberg.

James Young's chapter, "Mandating the National Memory of Catastrophe," is likewise concerned with the law as a means of mandating memory in the wake of the Holocaust. War crimes trials, however, play no role in his analysis. Young's interest is legislative, not judicial: He explores the law's role in mandating memory through official acts such as the creation of sites and days of commemoration. The story he tells is a fascinating one that invites comparison not only with Reichman's contribution but with Schafer's as well. In Schafer's reading of Balzac's *Colonel Chabert*, we encountered the flesh and blood survivor who found himself stripped of legal identity and citizenship as a result of being falsely reported as dead. In Young's essay, by contrast, statelessness and the absence of a legal identity appear not as a consequence of catastrophe but as a prelude to it. As Arendt observed in *The Origins of Totalitarianism*, rendering persons stateless is a crucial step to denying them all human rights, including the basic right to exist.[44] Young describes Israel's provocative effort to confer posthumous Israeli citizenship upon the perished Jews of Europe, a gesture that can be read as both an effort to redress the legal injustices occasioned by the Nuremberg laws of 1935 and those suffered by Balzac's doomed protagonist. Equally provocative is his account of the legislative genesis of Yom Hashoah, Israel's official day of Holocaust remembrance. In tracking the Knesset's attempt to pick a fitting day for Holocaust commemoration on the Jewish calendar—a calendar already crowded with the burden of the Jews' long, and often calamitous, history—Young reminds us that the protection and nurturing of memory are deeply political acts, as much prompted by a vision of the present as by a respect for the past.[45]

At the same time, Young ponders the matter of form, the particular mode of expression chosen by the Israeli parliament in its memorial gesture. If we tend to think of law as declarative and discursive, if not directly fulsome,[46] how do we make sense of the commemorative logic of Yom Hashoah, which honors memory through silence? Certainly, silence is a common form of paying respect to the dead, but as Young makes clear, in mandating silence over speech, Israeli law also deconstructs the logic of classic state-sponsored commemorative gestures. These earlier efforts, which often took the form of grand monumental and architectural statements, were designed to dictate the terms of collective memory and script the meaning of sacrifice to a national audience. Yom Hashoah, by contrast, resists this totalizing logic. By mandating collective silence, the Israeli law opens a space for memories to coalesce, collide, and collect. If Reichman describes a productive boredom that can give rise to memory, Young tells of a silence in which memory can speak. In its answer to catastrophe, law eschews the impulse to fix meaning. By choosing silence as its mandated response to catastrophe, the law both acknowledges the limitations of any gesture of amelioration and, at the same time, recognizes that only in the solemnity of silence can memory begin the project of repair.

Conclusion

Taken as a whole, what do these essays suggest about the relationship between law and catastrophe? We began by noting two conflicting master narratives: the divine and the liberal. In the former, catastrophe serves as an instrument of law; in the latter, law emerges out of the primordial chaos of catastrophe. In the liberal genealogy, catastrophe—conceptualized as the state of nature, the war of all against

TABLE 1.1
Regulative strategies of liberal legality

Strategy	Criminal Law (terrorism, genocidal regime)	Civil Law (corporate malfeasance such as industrial accident, collusion with genocidal regime)
Anticipation	Terrorism: domestic spying, torture and aggressive interrogation, preemptive military strike, etc. Genocidal regime: preemptive military strike	Promulgation and enforcement of regulatory norms (administrative law response: can have criminal law component) Assignment of corporate risk, doctrine of foreseeability (tort law response)
Prevention	Terrorism: indefinite preventive detention, torture, regime change Genocidal regime: regime change, war crimes trials (deterrence: weak justification)	Promulgation and enforcement of norms of safety and infrastructural development, expanded norms of corporate responsibility (administrative law response: can have criminal law component) Shifting of risk (tort law response), shareholder actions, divestment
Amelioration	War crimes: regime change, domestic and international trials, monuments, days of remembrance	FEMA (administrative law), strict liability (tort law), humanitarian relief

all—remains law's ultimate nightmare at the same time that it provides law's raison d'être. Law is indebted to catastrophe as it is vexed and troubled by it.

As a tool of maintaining social order through rules, liberal legality is predicated on certain basic divisions and regulative strategies. At its most basic, the world of harm is divided into criminal and civil wrongs, while law's regulative strategies involve techniques of anticipation, prevention, and amelioration. In its contact with catastrophe, we have seen law's struggle to assimilate extreme events into a regulative grid (Table 1.1).

Yet as our contributors have demonstrated, law's ordering strategies are neither conceptually neat nor altogether efficacious when it comes to regulating catastrophe. It is in the nature of catastrophe that it *demands* a legal response, at the same time that all ameliorative responses will be seen as insufficient and belated. Catastrophe demands that law's regulative strategies be devoted principally to anticipation and prevention, but as a limit condition, catastrophe has the power to distort and frustrate law's regulatory ambitions. In the domain of criminal law, draconian efforts to master the threat of catastrophic terrorist attacks threaten to erode liberal law's distinctive attractions as a tool of normative order. In the case of civil law, law's regulatory efforts reveal the instabilities in the distinction between prevention and amelioration, just as they threaten to exacerbate the very problems they are asked to relieve.

The relationship between catastrophe and liberal law is, in the end, at once constitutive and foundational, yet also liminal and unstable. Born out of catastrophic chaos, law strives to gain dominion over unmasterable calamity. Some, but not all, of these efforts may bear the traces of legal hubris and what Meyer describes as the colonizing impulse. We have also seen how contact with catastrophe also elicits law's humility, its respect of silence, and unscripted memory. Yet in its gestures of both hubris and humility, the law reveals its stress points in its continuing contacts with the specter of catastrophe. As the essays in this volume suggest, these stress points are both structural and contingent—built into the processes of legality but also amenable to change and reform. And while it will fall to future scholars of law and catastrophe to explore this balance between the structural and contingent more fully, we can say with certainty that these stress points will be tested severely in the years to come.

Notes

1. See, for example, Elizabeth Kolbert, *Field Notes from a Catastrophe: Man, Nature, and Climate Change* (New York: Bloomsbury, 2006).

2. See, for example, Laurie Garrett, *The Coming Plague: Newly Emerging Diseases in a World out of Balance* (New York: Farrar, Straus and Giroux, 1994).

3. Martin Rees, *Our Final Hour: A Scientist's Warning: How Terror, Error, and Environmental Disaster Threaten Humankind's Future in This Century—On Earth and Beyond* (New York: Basic Books, 2003).

4. See, for example, Anthony Cordesman, *Terrorism, Asymmetric Warfare, and Weapons of Mass Destruction: Defending the U.S. Homeland* (Westport, CT: Praeger, 2002); and Russell D. Howard and Reid L. Sawyer, eds., *Terrorism and Counterterrorism: Understanding the New Security Environment* (Guilford, CT: McGraw-Hill, 2003).

5. Voltaire, *The Complete Works of Voltaire*, Theodore Besterman et al., eds. (Geneva: Institut et Musée Voltaire, 1968).

6. Moishe Postone and Eric Santner, eds., *Catastrophe and Meaning: The Holocaust and the Twentieth Century* (Chicago: University of Chicago Press, 2003).

7. Elie Wiesel, *Night*, Stella Rodway, trans. (New York: Bantam Books, 1982), 32.

8. Primo Levi, *If This Is a Man*, Stuart Woolf, trans. (New York: Everyman's Library, 2000), 155.

9. See, for example, Nasser Hussain, *A Jurisprudence of Emergency: Colonialism and the Rule of Law* (Ann Arbor: University of Michigan Press, 2003).

10. See, for example, *The Selected Writings of Martin Luther* (New York: Anchor Books, 1959).

11. Amos 1:3–4.

12. Consider, for example, Pat Robertson's recent suggestion that God smote Ariel Sharon for "dividing God's land," http://www.cnn.com/2006/US/01/05/robertson.sharon/.

13. Thomas Hobbes, *Leviathan* (Cambridge: Cambridge University Press, 2002).

14. See Robert Cover, "Nomos and Narrative," in *Narrative, Violence, and the Law*, Martha Minow, Michael Ryan, and Austin Sarat, eds. (Ann Arbor: University of Michigan Press, 1995).

15. Judith Shklar, *The Faces of Injustice* (New Haven, CT: Yale University Press, 1990).

16. See, for example, Radley Balko's Fox News commentary: "Sept. 11 is no longer the most catastrophic failure of government in my lifetime. Its response to Hurricane Katrina is. Government at all levels, run by both parties, regardless of race, inexcusably failed to secure the safety of the people of New Orleans. The lesson here is not the failure of one party or the other. The lesson here is the failure of government." http://www.foxnews.com/story/0,2933,168732,00.html.

17. See, for example, the five-part series that ran in *The New Orleans Times–Picayune*, June 23–27, 2002, http://www.nola.com/hurricane/?/washingaway/.

18. Jean-Jacques Rousseau, *The Collected Writings of Rousseau*, vol. 11, Roger D. Masters and Christopher Kelly, eds. (Hanover, NH: University Press of New England, 1990).

19. *Black's Law Dictionary Eighth Edition*, Bryan A. Garner, ed. (St. Paul, MN: West Publishing, 2004), 1311.

20. For a recent critique of this process, see Cass Sunstein, *Laws of Fear: Beyond the Precautionary Principle* (Cambridge: Cambridge University Press, 2005).

21. For controversial answers in the affirmative, see John Yoo, *The Powers of War and Peace: The Constitution and Foreign Affairs after 9/11* (Chicago: University of Chicago Press, 2005).

22. See, for example, the pieces by Daryl Mundis, Ruth Wedgwood, and Harold Koh collected in "Agora: Military Commissions," *American Journal of International Law 96*, no. 2 (April 2002), 320–44.

23. See, for example, Alan Dershowitz, *Why Terrorism Works: Understanding the Threat, Responding to the Challenge* (New Haven, CT: Yale University Press, 2002).

24. See, for example, Bonnie Honig, "Bound by Law? Alien Rights, Administrative Discretion, and the Politics of Technicality: Lessons from Louis Post and the First Red Scare," in *The Limits of Law*, Austin Sarat, Lawrence Douglas, and Martha Merrill Umphrey, eds. (Palo Alto, CA: Stanford University Press, 2005).

25. Michael Ignatieff, *The Lesser Evil: Political Ethics in an Age of Terror* (Princeton, NJ: Princeton University Press, 2004).

26. Richard A. Posner, *Catastrophe: Risk and Response* (New York: Oxford University Press, 2004).

27. See, for example, Paul Shrivastava, *Bhopal: Anatomy of a Crisis* (Cambridge: Ballinger, 1987).

28. Guido Calabresi, *The Costs of Accidents: A Legal and Economic Analysis* (New Haven, CT: Yale University Press, 1970).

29. Posner, *Catastrophe: Risks and Response,* 9.

30. Ibid., 140–41.

31. See, for example, Gérard Prunier, *Darfur: The Ambiguous Genocide* (London: Hurst, 2005).

32. See, for example, Peter Evans, *Dependent Development: The Alliance of Multinational, State, and Local Capital in Brazil* (Princeton, NJ: Princeton University Press, 1979).

33. See Lawrence Douglas, *The Memory of Judgment: Making Law and History in the Trials of the Holocaust* (New Haven, CT: Yale University Press, 2001).

34. *Trial of the Major War Criminals before the International Military Tribunal,* vol. 1 (Nuremberg: International Military Tribunal, 1947), 99.

35. *Hannah Arendt–Karl Jaspers Correspondence 1926–1969,* Lotte Kohler and Hans Saner, eds., Robert Kimber and Rita Kimber, trans. (New York: Harcourt Brace Jovanovich, 1992), 54 (footnote omitted).

36. http://www.un.org/icty/legaldoc-e/index.htm; see specifically, Security Council Resolution 827, S/RES/827 (1993).

37. http://65.18.216.88/ENGLISH/Resolutions/955e.htm; see specifically, Security Council Resolution 955, S/RES/955 (1994).

38. William A. Schabas, *An Introduction to the International Criminal Court* (Cambridge: Cambridge University Press, 2001), 167.

39. See, for example, Martha Minow, *Between Vengeance and Forgiveness: Facing History after Genocide and Mass Violence* (Boston: Beacon Press, 1998).

40. See, for example, Douglas, *The Memory of Judgment.*

41. Hannah Arendt, *Eichmann in Jerusalem: A Report on the Banality of Evil* (New York: Penguin, 1994).

42. Rebecca West, *A Train of Powder* (New York: Viking Press, 1955), 3, 8, 9, 11.

43. See, for example, Hanna Yablonka, *The State of Israel vs. Adolf Eichmann,* Ora Cummings, trans. (New York: Schocken Books, 2004); and Douglas, *The Memory of Judgment.*

44. Hannah Arendt, *The Origins of Totalitarianism* (New York: Harcourt, Brace, 1951), 275.

45. See also James Young, *The Texture of Memory: Holocaust Memorials and Meaning* (New Haven, CT: Yale University Press, 1993); and *At Memory's Edge: After-Images of the Holocaust in Contemporary Art and Architecture* (New Haven, CT: Yale University Press, 2000).

46. See, for example, Peter Goodrich, *Legal Discourse: Studies in Linguistics, Rhetoric, and Legal Analysis* (London: Macmillan, 1987).

Catastrophe: Plowing Up
the Ground of Reason

LINDA ROSS MEYER

Catastrophe. The World Trade Center collapsing before our eyes, the San Francisco Earthquake, the Chicago Fire, Hurricane Katrina, the sinking of the Titanic, the engulfing of Pompeii, the Great Plague, nuclear war, the death of a child, the betrayal of a friend. These are all things that come to mind when one hears the word *catastrophe*. Whatever it means, it is something big and it is something bad.

From its Greek roots, *catastrophe* is literally an "overturning." The metaphor was originally understood as an overturning of earth, a "plowing." Catastrophe, then, is a plowing under of what has been sown. This crop failure, in the ancient world that this word described, is the defeat of all expectations, labors, hopes, plans. It means the farmer, despite all his care and hard work, will starve.

Catastrophe is also captured in English by similar metaphors: upsetting, overthrowing, upending. What is the domain of the catastrophic? What ground does it plow up?

One understanding of catastrophe is made possible by Judith Shklar in *Faces of Injustice*.[1] She argues that we-the-polity divide harmful events into either injustices or misfortunes. Injustices are wrongs done to one by others—overturnings of normative expectations. Misfortunes are all the "left over" bad things that happen that are not injustices. She argues that the boundary between injustice and misfortune is a cultural and political one: It shifts over time in response to changing expectations and new technologies of prevention, and it is also subject to conscious political change. In Shklar's view, we ought to trust the voices of those harmed and listen to their judgments about the boundaries between these categories. A recurring theme in her work is that many of the things we look at as misfortunes ought instead to be seen as injustices—especially "passive injustice"—failure to stand up for the rights of others and failure to alleviate the suffering of others. The traditional legal rule that there is no general duty to act to help others would be a prime example of the way in which, Shklar says, we have tolerated injustice by labeling it misfortune.

The way Shklar sees it, injustice is what overturns normative expectations. Misfortunes, by contrast, are failures or losses that one simply has to live (or die) with. They are a natural and expected part of life. Her argument is that we ought to *raise* our expectations and turn many of these misfortunes *into* normative overturnings. We should see them as injustices.

One suspects that Shklar would argue that the category of catastrophe is the same as the category of misfortune and just as subject to political reframing. So understood, many of the events we characterize as "catastrophes" should also be conceived as "injustices" if we acknowledge the polity's responsibility to take victims' views seriously and recognize affirmative duties to act to aid others.

In my view, catastrophes are in yet another category. They are not misfortunes in the sense of expected failures or "ordinary" losses that come with the business of life, in part because they are so large and overwhelming that no "business of life" can go on in the usual way. They are unexpected, normative overturnings, but they are not injustices either, because they challenge the very categories that form our judgment of just and unjust. Catastrophes can be remade into injustices as Shklar suggests, but that takes away their character as catastrophic.

So I would carve out a special place for catastrophe that is neither misfortune nor injustice. Catastrophes are normative overturnings, yet not injustices. They overturn our very faith in justice, plow up the ground itself. Catastrophes look like cosmic betrayals, not calculated risks. Catastrophes, whether local or international, are moments when we confront the limits of our normative world.

Catastrophes in this sense may or may not be a result of human agency. I would argue that the World Trade Center toppling down was catastrophic, despite being the result of human agency, because it brought into tragic relief the reality of global normative conflict against democratic capitalism. Characterizing it as crime would make (made) it easier to swallow, but the reason it was so profoundly disturbing was because of the depth of the antidemocratic and anti-American sentiments it revealed. We were suddenly confronted by a worldview that takes democracy and capitalism, our normative ground, to be evil.

Likewise, the Great Plague and the 1760 earthquake in Lisbon were catastrophes in part because the magnitude of the chaos and death they caused shook the world's belief in God and its faith in the social order. But catastrophes need not be on a grand social scale. A betrayal by a trusted spouse, friend, or child can cause one to lose faith in the normative world, to have the ground plowed up under one's feet.

Catastrophes, then, are not just bad luck, harmful events in the world, expected losses, or injustices but events that call into question our normative ground and cause radical normative disorientation.[2] They are events that cause us to feel "unheimlich"[3]—not at home in the world. Everything we took for granted is open to question. Everything we counted on is missing. We are plunged into an alien, senseless wasteland of the sort Kafka or T. S. Eliot describes.

Importantly, when we confront the failure of our normative ground, the edge of reason, we also experience our own lack of control. Where reason fails, so does prediction. The future is suddenly dark and uncertain. We here face our own limits, our own mortality. In Heidegger's terms, we experience "being-toward-death." In experiencing this finitude, we also experience our separation from others (the "they"). Because death is mine alone, I am I. We here run aground on our own limited and delimited individuality and, for the first time, know in a more than merely intellectual way that the "I" must always be experiencing death; there is always a path closing, a possibility gone, an end. This experience of senselessness, lack of control, and mortality that catastrophe plows up is profoundly uncomfortable and demands a response.

Our responses to catastrophe take three forms (at least). The first form is denial. We deny that the event is a challenge to our normative structures, and we reframe it as injustice, not catastrophe. This is our law's specialty. Law is constantly colonizing catastrophe, reframing it as injustice, expanding the bounds and jurisdiction of law, and consequently expanding the zone of human control and responsibility.

Tort provides a good analogy here for what is true of law more generally. In tort, liability turns on foreseeability, a term that tends to expand naturally with experience. As soon as something has happened, however unpredictable or unexpected, it becomes foreseeable.

Likewise, when any new form of misfortune occurs, legislatures rush to create new law to "fix" the problem, turning misfortune into injustice. In the wake of any disaster, one finds new legal regulation. In the case of the World Trade Center, we began from almost the outset to frame the event as a familiar form of injustice: a crime in need of retribution, coupled with negligence in preventing the attack—a glitch in the building firewall construction or airport safety precautions or intelligence-gathering protocols that can be improved upon next time. Predictably, we made new building codes, airport procedures, crimes, investigative tools, and so forth. Likewise, after Hurricane Katrina devastated New Orleans, we blamed FEMA, we blamed the city government of New Orleans, we blamed the Army Corps of Engineers; we responded with new procedures, new rules, new regulations and consoled ourselves that next time we would be ready, next time we would be in control. In these reframings, the catastrophic norm-overturning nature of the event hides itself, and catastrophe is reread as conventional injustice, à la Shklar. This response is the denial of catastrophe.

A second response is to "read" the catastrophe nihilistically. The nihilist acknowledges the normative challenge that the catastrophe represents and stays there. The normative ground is gone, anomie reigns, war and suicide are the only options. All that remains are the subjective claims of individual voices and the power that allows one voice, temporarily, to speak over the others.

In Shklar's urging that we should listen to how the victims of harm characterize that harm, because there is no absolute boundary between misfortune and injustice, there is a little hint of this nihilism. If she gives up her Rousseau-like optimism that there is a "natural" sense of justice that grounds and justifies these victims' claims, as it seems she must to hold that the misfortune–justice line is socially constructed, politics becomes the rule of the squeaky wheel or a civil war of competing demands. Every subjective complaint is as good as another. Justice cannot be picked out from jealousy.

But a third response is to see what we can from the perspective catastrophes open for us. We can stay in that very uncomfortable, unhomelike place and look. The catastrophic is a place from which we experience our finitude. Is there more to see there than death?

Catastrophe, the place where reason runs out, lets us see ourselves as unable to control the world. Catastrophe is what we didn't expect and couldn't predict. We see the world as outside our control and therefore other than us. We see that the normative ground, whatever it is and if it exists, is not of our making or within our control or even amenable to our reason. We stand before the world. The world is a given, beyond control. And so are we. We see that we did not make ourselves. For the first time, we have a perspective from which we can notice that our own urge to make sense of everything is itself a given, a grace, an unreasonable demand for reason. From the place where reason fails, we can see reason itself as our calling, a call from outside reason.

The insight that we are not self-made is also a moment in which we can see the basis of human responsibility. We notice that we are being receptive (responsive) to this world, this other, on the basis of limitations we must own without choosing. We have obligations we did not undertake, like the very basic obligation to reason (Kant's categorical imperative), but which nonetheless make us what we are. There is a fundamental injustice in all of this: We didn't choose ourselves, yet we are obligated. We are in our very essence something of a catastrophe— ungrounded, under the yoke of painful obligations we could not expect or imagine.

Acting responsibly (reasonably), then, is itself an act of faith or, as Kant puts it, an act done out of awe for the Law. This Law is the law of reason, a given that calls us, impels us to make sense of the world, impels even our knee-jerk first response: the making of injustice out of catastrophe.

This original obligation to make sense is itself ungrounded and unreasoned. The Kantian man of duty does his duty even though it hurts, regardless of conse- quences or happiness. He does it because that is the only way he can be responsible, the only way he can be. He responds on the basis of a given commitment to be reasonable, itself ungrounded. Within the moment of action, he cannot stop and consider the consequences or the cost or even the basis of the duty. The duty and his sense of awe for the duty are all.

This is heroism. Catastrophes create the conditions in which heroism can shine forth because it is only when the usual social system of rewards for virtue is swept away that one can see an act done purely from duty. Catastrophe breaks down the social expectations of justice, leaves us on the brink of nihilism, destroys our faith that we will reap the rewards of our virtuous and arduous sowing. Yet in catastro- phes, as we know, we find heroes. Heroes take on an impossible task and are will- ing to sacrifice in the faith that they are doing right, regardless of the consequences. From the brink of nihilism, catastrophe allows us, through the heroic, to snatch a glimpse at the core insight of human responsibility—that we are obligated to respond on the basis of an unreasonable demand that we be reasonable, even when there is no reason to do it.

Derrida calls this kind of phenomenon the aporia of justice. Positive law never captures justice, in part because language is always too general, rules too rigid, but also because one must act, in the end, without knowing everything. Both are prob- lems of finitude—the inability of categorical words to capture the infinite variety of experience and the inability of humans to gather all the information for all time. Justice is a call that we humans can never answer adequately because every deter- minable rule or system is necessarily inadequate to the particular: "justice exceeds law and calculation . . . the unpresentable exceeds the determinable."[4] Derrida would, perhaps, call catastrophe a moment in which we see this truth about law, and law is "deconstructed." Drucilla Cornell explains that "the deconstructibility of law is what allows for the possibility of transformation," a moment of "institutional humility before justice."[5] From this perspective, catastrophe allows us to glimpse a more problematic understanding of justice itself—a justice beyond reason, a justice that calls us but is never realized, a justice "between the lines" and just out of view.

This justice-as-aporia that impels the deconstructive project may be something like the givenness (or grace) of reason that I am saying is revealed when the "usual" normative framework fails. But it is not quite the same because Derrida's view might seem to imply that catastrophe (in the sense of normative dislocation) can itself be "created" (and therefore tamed and controlled) by deconstructive writing itself. Catastrophes would then become something we create in academic papers pointing out the deconstructibility of law. But one doesn't come to a point of normative dislocation in one's armchair, though one can "pretend" there, as Hume did, to doubt everything. Catastrophe can only really come upon one when all the reassuring reliability of everyday life can no longer contradict the mental exercise of normative deconstruction.

Derrida's justice-as-aporia is also only one possibility that may be opened when we confront catastrophe. His justice-as-aporia recognizes that justice cannot be captured for all time or in words. Justice-as-aporia is a kind of equity, always in conflict with law, always particular and fleeting. But even equity of this sort is only one possibility opened by catastrophe, for whatever we can or cannot say about it, we still know it when we see it and revere such Solomonic judgment. But the given at the edge of reason may not be only a still-familiar though inarticulate equity. It may not be at all. Or it may be grace—so foreign to us that we cannot even know it when we see it, but have to take it on faith. The moment of catastrophe may also be a moment of the deepest transformation—a transformation so complete and unsettling that we do not even recognize ourselves. This third response to catastrophe, then, allows for the possibility of experiencing grace.

To further illustrate these three approaches to catastrophe—denial, nihilism, and openness to grace—I call on an old story, that of Job.

Job is a very careful, responsible man, a worrier, and a planner. He is "careful not to do anything evil" (Job 1:1). He is so careful that he even offers sacrifices on behalf of his children the morning after they attend a feast, just in case "one of them might have sinned by insulting God unintentionally" (Job 1:5). Job ticks off his Leviticus checklist: He helped orphans and widows, he acted justly and fairly, he was faithful to his wife, he treated his servants fairly, he bought clothes for the poor, he never cheated even when he could get away with it, he never trusted in riches, never worshiped false gods, never exulted over his enemies' misfortunes, never turned away travelers, never concealed his sins. As a result, he "always expected to live a long life and to die at home in comfort" (29:18).

But instead, "everything I fear and dread comes true." His livestock are stolen, his children die, his body is so riddled with hideous disease that his friends don't recognize him, and his wife avoids him. His future, "roots and stock," is destroyed. He suffers catastrophe, an overturning of everything he has hoped for and planned.

Job doesn't understand it; he has tried his utmost to follow the law, and yet he gets no justice. His good deeds, the tremendous burden and struggle to be vigilant and responsible, are all for nothing. He complains to God "as soon as I sin, I'm in trouble with you, but when I do right, I get no credit" (10:15). "Why is man so important to you? . . . Won't you look away long enough for me to swallow my spit?" (7:17–19).

Until this point, Job has been a man of the law, perhaps even "the" reasonable man of tort law, foreseeing and averting harm, buying insurance, following the law meticulously, and paying his taxes on time. His world is grounded in law, and it seems to him that God has made some mistake. His first response, then, is to reframe his plight as injustice. He is a man of the law in more than one sense, for he has himself been an esteemed and honest judge in his town, and he thinks and speaks in legal terms, asking "why doesn't God set a time for judging, a day of justice for those who serve him?" (29:11–25; 24:1). He asks again and again for justice, for a chance to "argue my case" with God (13:3). He asserts, "I am ready to risk my life. I've lost all hope, so what if God kills me? I am going to state my case to him . . . I am ready to state my case, because I know I am in the right" (13:14–15, 18). But how does one bring God to account? "Should I try force? Try force on God? Should I take him to court? Who would make him go?" (9:19).

He has four faithful friends who come to commiserate with him, and together they try to sort out why he is suffering such unmerited catastrophe. It is a legal debate, a courtlike discussion in the style of adversary argument. Three of Job's friends have trouble believing that God would allow good men to suffer so acutely, and therefore, they reason, Job must be guilty of some fault. The friends, then, take the first approach to catastrophe: denial. They try to reframe the catastrophe as a just punishment for Job's injustice. Eliphaz argues, "evil does not grow in the soil, nor does trouble grow out of the ground. No! Man brings trouble on himself, as surely as sparks fly up from a fire" (5:6–7). Bildad says, "your children must have sinned against God, and so he punished them as they deserved" (8:4). And Zophar admonishes, "God is punishing you less than you deserve" (11:6).

Job is impatient with them. "I will never say that you men are right," he maintains. "I will insist on my innocence to my dying day" (27:5). But the failure of the possibility of framing his suffering in the usual category as injustice means law has failed. This failure of law leaves Job with the second response, nihilism. Job laments, "I am innocent, but I no longer care. I am sick of living. Nothing matters; innocent or guilty, God will destroy us. When an innocent man suddenly dies, God laughs. God gave the world to the wicked. He made all the judges blind. And if God didn't do it, who did?" (9:21–24). Life no longer makes sense, "my plans have failed; my hope is gone" (17:11), chaos reigns, and Job curses the day of his birth and longs only for the relative peace of death in a land of "darkness, shadows, and disorder" (10:22).

Finally, his three older friends give up arguing with him, and his youngest friend, Elihu, speaks his mind. He is angry that the older men have given up defending God. He repeats his conviction that God is just, but he stresses that "if God decided to do nothing at all, no one could criticize him" (34:29). He argues that "those who are godless keep on being angry, and even when punished, they don't pray for help . . . But God teaches men through suffering and uses distress to open their eyes . . . Be careful not to turn to evil; your suffering was sent to keep you from it" (36:13, 15, 21). Elihu gives us the first hint that there may be a third response: a response of openness to catastrophe.

Finally, God himself comes to speak to Job. He has granted Job's request for a hearing, but He makes it clear that He is not there to answer Job's accusations. Instead, He is plaintiff, not defendant, and He demands an answer from Job: "Stand up now like a man and answer the questions I ask you" (38:3). This move is critical: God is not called by the other; He does the calling. He is active, not reactive, not responding. It is instead Job who is called to respond, able to respond, required to respond, responsible.

God first points out that Job is not to be his own judge—he is unqualified to have jurisdiction here. "Were you there when I made the world? . . . Have you been to the springs in the depths of the sea? Have you walked on the floor of the ocean?" (38:4, 16).

Second, God points out that the laws Job relies on cannot bind God. Following God's law gives him no settled expectations, no right to demand anything. "Do you know the laws that govern the skies, and can you make them apply to the earth? . . . Can you shout orders to the clouds? . . . Will a wild ox work for you? . . . Can you catch Leviathan with a fishhook . . . Will he make an agreement with you and promise to serve you forever? Will you tie him like a pet bird, like something to amuse your servant girls? Will fishermen bargain over him?" (38:33–34; 39:9; 41:1, 4–6).

Third, God points out that other creatures do not live in concern and fear for the future and that, perhaps, Job has been too obsessive about his security and his calculated planning. The creature of God is not to act like an actuary. The ostrich "leaves her eggs on the ground for the heat in the soil to warm them. She is unaware that a foot may crush them or a wild animal break them. She acts as if the eggs were not hers, and is unconcerned that her efforts were wasted. It was I who made her foolish and did not give her wisdom. But when she begins to run, she can laugh at any horse and rider" (39:14–18). The horses "rush into battle with all their strength. They do not know the meaning of fear, and no sword can turn them back" (39:21–22).

Job responds twice. First, he says, "What can I answer? I will not try to say anything else. I have already said more than I should." But God is not satisfied and calls on him again to answer "like a man." A man stands and answers to the call of the other. A man must respond, be responsible. Job's second response is this: "You ask how I dare question your wisdom when I am so very ignorant. I talked about things I did not understand, about marvels too great for me to know . . . In the past I knew only what others had told me, but now I have seen you with my own eyes. So I am ashamed of all I have said and repent in dust and ashes" (42:3, 5–6).

Job's answer, then, is to acknowledge his finitude and repent. He was wrong to call God to answer. God is the other, the unanswerable, the not-called-to-respond. He was wrong to make God into a security dispenser. As Elihu said, suffering has brought him wisdom. But more, he has seen God with his own eyes. Nothing he has been told before has prepared him for this. All his preconceptions are gone; all is changed. Job is now open to a completely unforeseen grace.

God restores Job's fortunes then and punishes his first three friends, "because you did not speak the truth about me, the way my servant Job did" (42:7). He asks Job, then, to pray for his friends, and then He "answers Job's prayer." He does not answer in court, but in prayer, an entirely different venue. He answers with grace, not justice.

Job, it seems, was right to recognize that God did not do (human) justice but wrong to complain of it. Job's friends, who tried to assimilate God to human justice and expectation, made God "angry." They "did not tell the truth" about God. They made the mistake of creating a mechanical God, a God of Reason, a God within human control, a cosmos controlled by and accessible to human understanding. In doing so, they thereby refuse to see the "other" as other at all. They deny the normative overturning that is catastrophe and reframe it in terms of justice to make catastrophe fit the mold of reason. The problem is, they believe, that Job simply hasn't figured out what went wrong, but once his friends uncover the unknown sin, all will be explained.

This is the response of denial, which treats catastrophe as a failure of foresight to be factored into an improved predictive mechanism for the future. Catastrophe gradually loses its character as other and is continually colonized by reason, as a kind of ever-expanding foreseeability. Catastrophe is unable, then, to arrest us, cause us to pause in terror, wonder, or awe. It cannot reach us, and therefore it can neither destroy nor transform us. When Job's friends' final acknowledgment of the undeserved nature of Job's catastrophe destroys even their best attempts to reframe it, the three friends fall silent, unable to defend God further. They, like Job, sink into a lassitude, boredom, and hopelessness that is the second response to catastrophe—the "last man's" nihilism.[6]

Job must recognize that there is catastrophe in the world, and good people suffer. But the other is not bound by law, cannot be controlled or ordered about by man. The other is other, beyond Job's reach, knowledge, ken. Indeed, in Job's encounter with God, he uncovers the ground of responsibility. He must respond to the call of justice, this new justice that he cannot understand or rationalize, but it need not respond to him. He is responsible. God, however, is not responsible, but other. Job must give up justifying himself to God by the law.

Finally, Job acknowledges his finitude and repents, accepting what God has done. He gives up his claim, and he answers the call of the other in the face of catastrophe, ready to give up his life "in dust and ashes" or to receive it again as an undeserved gift, his birthday no longer accursed to him. His death is all that is left to him, all he can respond with; he is struck by catastrophe down to the nub of his self. He owns only his finitude. His true commitment to God can only shine forth in the moment of catastrophe, when he has nothing left to gain from it. He is then truly responding on the basis of duty alone; he is then heroic. Only then can we (doubting devils) see Job's commitment as completely pure and un-self-interested.

Only then can he "repent" in a completely pure and un-self-interested way, for he hates even his own life. The Kantian experiment of true duty is what the Devil demands and that of which Job ultimately proves worthy. In the nearness to God that catastrophe brings, Job's near-nihilism is at last replaced with awe, the acceptance of possibility, gift, and grace.

If we take the first response to catastrophe, and always restructure the story we tell about catastrophe to turn catastrophe into injustice, we regain our feet, but we lose an opportunity to see ourselves and the world differently. We lose the possibility for heroism and compassion and the possibility of experiencing grace.

Heroes don't usually appear in stories of justice and injustice. The person who aids another in a catastrophic setting and without any hope of reward is not a hero, but someone who is merely doing what he or she ought to do. And in fact, the law may discover that the hero should have done so earlier or differently and is still guilty of injustice. Heroism, and the dignity we ascribe to it, disappears in the hindsight bickering over exactly what course of action would have been most effective.[7] From the perspective of reason, we can no longer see the relevance of devotion to duty (which, after all, is not within reason's grasp) and can only judge the act by its conformity to a perfectly rational calculation. Hence, reframing catastrophe as injustice hides the heroes.

Second, we owe victims of catastrophe turned to injustice their just deserts. There is no more compassion, only compensation. Outpourings of money for destitute persons are only what is right, not what is good. These gifts do not create new connections forged of compassion and gratitude, or express solidarity and community,[8] but become only a grim and necessary satisfaction of tort claims, a sort of compensation that usually results in resentment and ends relationships rather than creating them. Justice replaces a gift economy with a market economy and destroys the community that gift exchanges may create.

Of course, one would always rather be owed than be given because the former generates no obligations (or gratitude or return), whereas the latter does. And in our market culture, receiving "charity" is demeaning. *Goldberg v. Kelly* (397 U.S. 254, 1970) famously changed a gift frame into a justice frame; recipients of welfare benefits are "entitled" to rather than merely given them. There seems something ennobling about having entitlements rather than gifts. It seems to bespeak equality between the giver and receiver. Yet something is still lost, even if something else is gained. The benefit turned to entitlement is no longer a sign of care or concern, but just "what is owed." As a result, the relationship easily comes to seem adversarial. There is no reason, anymore, for the giver to err on the side of generosity but a tendency to give as little as necessary, as little as is "due."[9] Likewise, the receiver now has standing to complain that the benefit is too little, instead of experiencing the care and solidarity the gift shows or being anxious to live up to the obligation and trust it imposes.

Third, if we respond to catastrophe with denial, our normative ground is never challenged but is instead reinforced by refusing to allow catastrophe to let us stand before the edges of our reality. No radical new possibilities are opened. We experience no humility or self-interrogation or transformative potential. So not only do we give no grace, but we also receive none.

What would transformative grace look like? To make it more concrete, I return to the examples given earlier of the three kinds of catastrophe: those caused by human agency, those involving widespread destruction, and those that are personal.

Seen in the conventional framework of injustice, the September 11 attacks lead us to set about demanding justice and reinforcing values of capitalism and liberal democracy in the world. But we could also face up to the disquieting, catastrophic aspect of those attacks, the thing that made them so much more disconcerting than the Oklahoma City bombing, for example. We can stand fearlessly at the edge of the chaos and confront the depth of our enemy's hatred of us, along with the possibility that capitalism and liberal democracy are not the solid normative ground we thought they were. Maybe there are other truths outside our ken that we could hear if we listen. Maybe our enemies have a role to play in our own national transformation. Maybe we can see the "other" across the divide.

The massive social dislocation and destruction wreaked by the Great Plague were seen, from an injustice perspective, as a judgment from God, requiring inquisitions and a redoubling and centralizing of legal authority to quell evil.[10] But maybe the catastrophe also allowed some to see the normative framework itself as flawed. Maybe catastrophe showed that God shouldn't be understood as a reward dispenser, and maybe the Reformation and the Enlightenment, in their different ways, could be understood as flowing from a response to catastrophe that fundamentally transformed the normative ground.

Abuse, the end of a marriage, and the death of a child are also places where catastrophe tears away the normative landlines of one's life. That moment can either be understood as an injustice (as it often is in divorce, criminal law, or tort litigation) or as an opportunity for mourning, listening in the void, being open to something new, unsettling, and "other." The story of Job tells of this kind of transformation.

A more secular example of personal transformative grace is found in Annie Proulx's novel *The Shipping News*.[11] Quoyle, a bland doormat of a character who is the passive, stunted victim of an unfaithful, child-selling spouse and a cruel boss, magically uncoils in the warmth of cold Newfoundland and discovers new bonds. Logic, Proulx's wordplay makes you see, has nothing to do with grace. Quoyle's journey away from all that is *terra cognita* to the literal unknown of New-found-land gives him a new set of possibilities—possibilities he never before knew existed. He becomes a writer, he learns how to boat, he finds moorings, he finds love, he even becomes a hero. In the course of the novel, he is transformed from a feeble outcast to a strong, warm, integral member of an irrational, generous community. Proulx uses all the symbolism of resurrection throughout her novel to underscore Quoyle's transformative rebirth. From the edge of reason, the edge of the world, one can see possibilities beyond reason.

In sum, when a catastrophe challenges the normative ground from which we make sense of the world, we are predisposed (by our own rational nature) to "read" that catastrophe as an injustice. We don't want to remain in that morally bewildering and lost place, and if we do, we risk nihilism. But moving too quickly to reframe the pain as injustice rather than catastrophe may obscure and indeed eliminate the opportunity to see heroism and compassion and to experience the illumination or transformation that Derrida's transcendent "mystical" justice or Martin Luther's radical grace or Kant's "awe" or Heidegger's "call" of care always urges on us.

Notes

With thanks to Austin Sarat, Martha Umphrey, Nasser Hussain, Lawrence Douglas, Roger Berkowitz, and the participants in the workshop on Catastrophe and Law at the Amherst Department of Law, Jurisprudence, and Social Thought for inviting me to give this paper and challenging, redirecting, and honing my thoughts. Many thanks also to Jeff, Dan, and Amy Meyer, Sandy Meiklejohn, Melissa and Doug Logan, Steve Gilles, Steve Latham, and Shai Lavi for catastrophic conversations and to Emilie Waters for brilliant research assistance.

1. Judith Shklar, *Faces of Injustice* (New Haven, CT: Yale University Press, 1992).

2. Friedrich Nietzsche, *The Genealogy of Morals*, Walter Kaufmann, trans. (New York: Penguin, 1969), 68: "What really arouses indignation against suffering is not suffering as such but the senselessness of suffering."

3. Martin Heidegger, *Being and Time*, John Macquarrie and Edward Robinson, trans. (New York: Harper & Row, 1962), 232.

4. Jacques Derrida, "Force of Law: The Mystical Foundation of Authority," *Cardozo Law Review 11* (1990), 971.

5. Drucilla Cornell, "The Violence of the Masquerade," *Cardozo Law Review 11* (1990), 1060–62.

6. Friedrich Nietzsche, "Thus Spoke Zarathustra," in *The Philosophy of Nietzsche*, Thomas Common, trans. (New York: Modern Library Edition, 1950). In Section 2, the outworn philosophy advocates "sleep, the lord of the virtues."

7. With one slight twist. The lawyers or judges who bring about justice in the wake of social norm breakdown then are the heroes of the story. They become the heroes who are doing their duty without regard to selfish motives (e.g., Atticus Finch). Yet, when a justice perspective is brought to bear on them, we see only what more they could have done, as in a malpractice case, or how the system ought to be improved.

8. Lewis Hyde distinguishes gift and market economies and describes how the gift circle of potlatch ceremonies connected Native American tribes to each other and cemented relationships. *The Gift: Imagination and the Erotic Life of Property* (New York: Random House, 1983).

9. Philippe Nonet shows how workers' compensation boards in California became less generous when their jobs became framed by legal process and legal ordering. *Administrative Justice: Advocacy and Change in a Government Agency* (Hartford, CT: Russell Sage Foundation, 1969).

10. Robert C. Palmer, *English Law in the Age of the Black Death, 1248–1381* (Chapel Hill: North Carolina University Press, 1993).

11. Annie Proulx, *The Shipping News* (New York: Scribner's, 1991).

Catastrophes and Humanitarian Corporate Responsibility: A Conceptual Critique

RONEN SHAMIR

In recent years, new demands for legalizing and for establishing a right to humanitarian intervention have been firmly placed on the international agenda. Still, most attention is given to the actions of states and suprastate bodies that are expected to intervene in protecting the human rights of victim populations. In this chapter, I argue that the duties of multinational corporations (MNCs) that operate in areas where humanitarian intervention is considered should also be addressed. Premised on the idea that MNCs are in an increasingly strategic position to anticipate, prevent, and contribute to the alleviation of catastrophes in many host countries, we should begin thinking about creating a framework for sanctioning corporate humanitarian intervention—mediated through their duties to protect, anticipate, and prevent violations of human rights—even before considering armed humanitarian intervention by foreign countries and international organizations. Yet beyond the normative engine that drives this chapter, my primary theoretical and empirical effort is to show how contemporary corporate practices are geared away from assuming binding legal obligations. I show this by looking at prevailing notions of corporate social responsibility (CSR) in relation to humanitarianism. I try to show that the basic disposition seems to be away from binding, enforceable, and sanctioned legal obligations and toward reliance on voluntarism, altruism, and self-regulation.

Fusing analytical and normative dimensions, the discussion develops as follows. I first introduce the field of corporate social responsibility, point to some of its most basic characteristics, and highlight some of the conceptual debates about the desired meaning and application of the term. I next consider some basic principles of humanitarianism and point to some leading trends and debates concerning its trajectory. In the third section, I argue that through their respective locations

within the neoliberal conceptual framework of the relationship among states, markets, and civil society, we can appreciate the structural homology between prevalent practices of humanitarianism and contemporary corporate-inspired notions
of social responsibility. The fourth and fifth sections consider two concrete cases
in which corporate social responsibility and humanitarianism intermesh or po-
tentially intermesh. I conclude with some remarks concerning future potential
obligations of corporations in times of catastrophes.

On Corporate Social Responsibility

Talisman Energy Inc. (formerly, British Petroleum Canada) is a multinational oil
and gas producer and a natural gas supplier incorporated in Canada. It is among
the top sixty companies on the Toronto Stock Exchange and is also traded on the
New York Stock Exchange. In its annual report for 2002, Talisman reported a net
income of Can$524 million and a total assets value of Can$11,594 million. On its
Web site, alongside its annual report, Talisman also published its corporate responsibility report for 2002, which stated the following:

> At Talisman, corporate social responsibility means conducting activities in an economi-
> cally, socially and environmentally responsible manner. It also includes working together
> with stakeholder groups to identify constructive solutions to shared problems. We be-
> lieve that our operations bring direct benefits to the communities in which we work,
> including the creation of jobs, expansion of local infrastructure and support of commu-
> nity projects that create opportunities for a better future. As a responsible business, we
> also believe it is our duty to observe and promote ethical business practices and advocate
> respect and tolerance by and for all people. In 2002, we took steps to further integrate
> corporate responsibility activities and objectives within our corporate governance and
> management systems and expanded our reporting by enhancing disclosure regarding
> environmental and economic transparency. The 2002 Corporate Responsibility Report
> details these developments and provides a broad overview of our social, environmen-
> tal and economic activities in each of our principal geographic regions (http://www
> .talisman-energy.com).

Talisman's thirty-six-page social responsibility report covered issues such as human rights, community programs, ethical business conduct, employee relations,
environmental audits, waste management issues, and transparency practices. The
report was certified by PricewaterhouseCoopers (PWC), a global auditing firm.
The report was structured along the suggested principles of the Global Reporting

Initiative (GRI) and followed the Triple Bottom Line model (economic, environmental, and social reporting) that in recent years has emerged as a blueprint for "responsible" reporting. Among its human rights activities, Talisman reported that it established a security policy to deal with the conduct of national defense forces protecting corporate facilities. It reported that this policy emerged out of its experience in Sudan, where it tried to prevent the use of its oil field facilities for "non-defensive purposes." Talisman also reported on its human rights training initiatives and projects in Sudan and Colombia as well as on its specific management efforts to promote peace in these two countries.

Talisman was thus directly involved in social, political, and diplomatic activities that until recently were assumed to belong with governments, international institutions, and nongovernmental organizations. The activities of Talisman will occupy us in latter parts of this chapter. For now, suffice it to say that as this example shows, the ideas and practices of corporate social responsibility have in recent years been rapidly institutionalized and professionalized. Corporate social responsibility is increasingly incorporated into managerial systems and organizational culture practices, exponentially developed and practiced by experts of various sorts (accountants, auditors, consultancy firms, public relations firms, lawyers, and other new specialists in "social responsibility"), and studied in business management schools.

Serious debates about the substantive merit of CSR relative to its role as a marketing and image-management device are common.[1] Nevertheless, it seems quite certain that the evolution of CSR into a professional area of expertise and into a field of action that is widely employed by corporations has profound effects on the very meaning and scope of social responsibility. Indeed, this assigned meaning is the major currency that is negotiated in the CSR field. At one end of the spectrum are players who associate the term *responsibility* with an ever-increasing set of moral duties and legal obligations. These players—sometimes referred to as "confrontational" actors—try to invest the idea of CSR with binding and enforceable rules. They envision it as a set of regulated structures of corporate governance operating at the national and transnational levels.[2] At the other end of the spectrum are corporations and a host of other affiliated players who associate the concept of CSR with voluntary, nonenforceable regulatory practices. Thus, corporate activities currently encompass a variety of declarations and commitments, including "codes of conduct," "mission statements," and "social auditing schemes," all designed to display corporate acceptance of the general idea that they do bear social responsibilities. Yet the most distinctive common denominator of all these corporate-based notions of social responsibility is the voluntary—at times altruistic and at times utilitarian—meaning of the term.

Attempts to move CSR into the enforceable domain of formal regulation have made little headway. For example, the European Commission recently rejected proposals to adopt a regulatory approach that would have subjected corporations to mandatory social and environmental reporting. The EC emphasized the "voluntary nature of CSR" and clarified that it did not intend to impose responsible behavior on companies by means of compulsory regulation.[3] Likewise, attempts to subject MNCs to the jurisdiction of the recently established International Criminal Court on the grounds that MNCs should be held liable for violations of international law had also been aborted.[4] Attempts to curb corporate power through law also involve the mobilization of the "developed" legal systems of rich countries to police and sanction corporate practices that take place in impoverished and exploited countries. A case in point concerns the attempt to subject MNCs to U.S. federal jurisdiction by invoking the Alien Tort Claims Act.[5] As we shall see, the future of this route remains unclear, but it already meets fierce opposition from corporations, business organizations, and significant elements in the American legal and political establishment.

Another example of corporate safeguarding against legalizing social responsibilities was evident in the reaction of multinational pharmaceutical companies to charges that the pricing policies of patented HIV drugs had constituted a serious impediment to the ability of the South African government to combat the AIDS epidemic and to the population's access to drugs. Facing mounting popular pressures, many pharmaceutical companies were quick to announce a host of philanthropic campaigns for the free distribution of drugs to selected populations in South Africa and a variety of socially responsible projects. For example, Merck announced that it was lowering its prices for the company's two antiretroviral medicines used to treat HIV infections. "At these new prices," the company's notice said, "Merck will not profit from the sale of these medicines in the developing world." Merck stated that its goal was to "spur efforts to accelerate access to these life-saving medications" and that it was Merck's third major initiative in less than a year regarding access to HIV/AIDS medicine in Africa.[6] Likewise, Pfizer offered to give away Diflucan, an expensive AIDS drug, to poor South Africans as part of its policy of responding to the "unmet medical need" in the country.[7]

Yet at the very same time, these companies launched a fierce legal struggle against a new South African law that allowed for the importation of significantly cheaper generic HIV drugs. Underlying the South African dispute was the attempt of the pharmaceutical industry to preserve the boundary separating *legally protected* business interests from *voluntary* practices of business social responsibility. The strategy pursued by the pharmaceutical companies was to disengage the dispute from the question of cost and to ground it in principled constitutional questions. The companies thus argued that the dispute was not about the relationship between the cost and the unavailability of medicines to large portions of the population. Rather, it was over the constitutionality of the legislative means chosen by the government—means that amounted to a serious infringement of their property rights. It was only when the strategy failed, due to a well-organized activists' campaign inside and outside the court, that the pharmaceutical companies withdrew their case.[8]

The CSR field as it currently operates is firmly based on a voluntary and self-regulatory orientation. Corporations and trade associations try to ensure that CSR will remain outside the formal domain of law. Above and beyond the variance among social responsibility schemes and displays, they all share a voluntary approach, and they are all based on principles of nonenforceability. The grounding of this particular meaning of responsibility as voluntary and unenforceable is aided and legitimized by theories, studies, professional conceptions, and policymaking approaches that, however diverse in purpose and aspirations, agree on the fundamental failure of the state-centered "command and control" mode of regulation. New models of regulation are thus based on soft-law approaches, metaregulation, responsive regulation, private regulation, or self-regulation, all reconfiguring the regulative role of the state and steering it toward, at best, governing at a distance. The mechanics of the process are based on the decentering of the regulative role of the state, reconfiguring it as a facilitator of a multistakeholder approach to regulation, and involving civic and commercial players alongside state-based organs and international bodies. The governing technologies that are launched, in turn, are based on dialogical persuasion, willing cooperation, internalized commitment, and even moral sensibilities. This logic of responsibilizing nonstate actors to willingly assume tasks formerly performed by state organs—often coded as "governance"—neatly fits the voluntary approach to CSR. Within this paradigm, the corporation is best driven to perform social responsibilities when it is exposed to consumer and investor demands; engages in dialogue with civic groups, communities, and private authorities; and learns to transform CSR into a corporate asset (e.g., reputation, employee satisfaction, competitive advantage, or risk-management strategy).

On Humanitarianism

For purposes of the ensuing discussion, catastrophes are conceptualized here

> not only as grand spectacles of destruction, devastation, and suffering, blocking the way to progress; they also promote and enhance new global orders of governance. The unmanageability of some natural and human-made disasters and the pain they inflict on a large number of civilians send shockwaves throughout the social, economic and political fabric at the local, regional, national, and transnational levels. Responses are called for. A sense of moral emergency is aroused. Responsibilities are negotiated and assigned. Sovereignty is challenged. Root causes are being explored. Solutions are offered. A host of experts and volunteers, governmental and non-governmental organizations, civilians and armed forces, are spread over the catastrophic terrain. Communities are targeted for reconstruction. Above all, the very understanding of and reaction to catastrophes activate a moral test in which changing perceptions of moral responsibility are being played out.[9]

Following this conceptualization, the argument here is that basic (albeit contested) principles of humanitarianism—namely, wide-scale assistance efforts targeting victims of catastrophes—are by and large based on the compassion and voluntary readiness of free-willing agents to act in the face of distant suffering. To a large extent, notwithstanding the dictates of international humanitarian law at times of war, humanitarianism matured and developed outside the domain of law. The duty to act is moral, and even when this morality is coupled with utilitarian reasoning, it still insists on its autonomy from binding legal duties. To some extent, legal codification is treated as if it contaminates and robs that morality of its political disinterestedness.

The positioning of humanitarianism outside the domain of law is also related to the fact that both national and international legislation are by and large state centered and aim at the protection of national sovereignty. Humanitarian intervention, even when it is carried out with the consent of the host government (let alone when it isn't), necessarily infringes on sovereignty, real and symbolic. In this sense, the perceived moral duty of humanitarian intervention works against states. In contrast to social rights that connect human and citizen rights, humanitarianism works to "separate the status of being human from the status of citizenship," in fact allowing for a new universalizing principle to set the conditions for social action.[10] Partially to alleviate this inherent tension, the dominant blueprint of humanitarian action (at least until recent years) has been characterized by strong commitments to principles of impartiality, neutrality, and confidentiality, most notably demonstrated in the model operations of the International Committee of the Red Cross (ICRC). That is, humanitarianism is positioned not simply outside law but also outside politics, abstaining from coupling pity with justice, relief with obligatory responsibility, and aid with the search for the root causes behind a catastrophe.

Not least significant is the fact that humanitarianism is squarely positioned on the side of "relief" in the commonly used relief–development distinction. Aspiring to conceptualize two distinct ways of addressing human want, relief is generally perceived as the short-term or emergency supply of vital commodities and services to victims of a catastrophe. Development is understood as a longer term investment process that enables chronically marginalized communities to enjoy enhanced physical and material prospects for self-reliance. Development is assumed to expand economic productivity and to affect social organization and political power, whereas relief is designed to alleviate immediate and tangible suffering.[11] Development is thus inherently tied to broader economic and political concerns (planning, investment, policies, international contracting, joint ventures, etc.), whereas relief is presumably secured precisely through its disengagement from such broader aspects of the political economy. Relief, and the humanitarian gesture within it, is thus also structurally located as the performance of an imagined universal civil society: spontaneous, voluntary, morally inspired, solidarity oriented, and politically neutral. Unlike development efforts that imply political choices, writes Senarclens, charity-based humanitarian relief impartially "mobilizes emotions rather than political reflection."[12]

Indeed, the tremendously vigorous and rapid expansion of humanitarianism in the last quarter of the twentieth century is a vital element in the overall ascendance of neoliberalism as a dominant ideological regime. The neoliberal framework is not only about free trade, self-regulation of markets, and the privatization of state services. It is also about a perceived role for civil society as a generator of social activities rooted in notions of spontaneity, solidarity, and altruism. Conjuring images of charity events, civic associations, and good citizenship, civil society has become a "channel for the diffusion of neo liberal norms."[13] Projected as the embodiment of public moral sentiments that bind humans together, civil society is perceived as a locus of spontaneous action based on individual initiative and of collective organization on a voluntary basis apart from state, law, and commercial interests. Conceptualized by neoliberals in terms of a social laissez-faire doctrine, the concept of civil society is thus recruited as "a substitute for the state, taking over functions like welfare or humanitarian assistance."[14] In concrete terms, this also means that the ascendance of humanitarianism is coupled with the proliferation of nongovernmental organizations (voluntary "civil society" associations) as primary agents of humanitarianism. Thus, the self-regulation of markets on the one hand and spontaneous civil action toward victims of catastrophes on the other hand have become the double bind of a neoliberal framework that works to depoliticize and hence situate both CSR and humanitarianism outside the domain of law and within the firm grip of the domain of civic virtue.

However, not unlike struggles taking place within the field of CSR, there has also been an enhanced debate about the prospects of legalizing and politicizing humanitarian interventions in recent years. Bernard Kouchner, founder of Médicins Sans Frontières (MSF), has been a pioneer in calling for transcending the heretofore apolitical humanitarian model and for bringing together humanitarianism and politics. MSF, winner of the 1999 Nobel prize for peace, has begun to integrate its commitment to providing medical care with an advocacy approach that emphasizes the duty of "bearing witness and speaking out": speaking out against human rights abuses and violations of international humanitarian law that its teams witness while providing medical relief. Specifically, MSF defines its mission as "rebellious humanitarianism"; namely, it does not see its mission merely in terms of "doing good." Rather, rebellious humanitarianism means political engagement in the sense of taking an active role in the actual definition or redefinition of situations as disastrous or catastrophic, defining situations in terms of human rights abuses, and defining situations in terms of those responsible.

The approach advocated by MSF culminates in the idea of pressing the United Nations and other global organizations to recognize a right to humanitarian intervention that would legalize a limit to national sovereignty and, when needed, legitimize forced intervention.[15] No doubt, the idea of overriding state sovereignty in defense of human rights marks a conceptual break with a state-centered approach and represents a move toward conceiving armed humanitarian intervention as international law enforcement. Michael Walzer articulates the emerging new conception of obligatory humanitarian intervention as follows: Humanitarian intervention has so far been essentially based on the paradigm of philanthropy, thereby voluntarily addressing the needs of victims. However, a shift of paradigm is needed, mediated through the language of human rights, stressing the rights (rather than needs) of victims to have their human rights protected and enforced.[16]

However, the legalization of armed humanitarian intervention opens up a plethora of new dilemmas. Humanitarianism can be appropriated as the ideology of rich states, enhancing their world position and providing legitimacy for military interventions prompted by strategic geopolitical considerations.[17] As Mamdani reminds us, there can be no such thing as an unambiguous humanitarian intervention.[18] Every intervention in the future will serve a complex of interests as much as every imperial intervention in the past also claimed to be humanitarian. Calling an intervention "humanitarian," in short, cannot strip it of its politics. Still, those in favor of legalizing humanitarian intervention are doing so precisely in the name of coupling humanitarianism and politics, albeit arguing for a higher form of politics—namely, that of international law enforcement sanctioned by accountable international bodies.[19]

Not unrelated to the question of politicizing and legalizing humanitarianism, there has been a recent effort to transcend the relief–development divide and to free humanitarianism from its firm positioning as a relief platform alone. For example, it has been argued that relief operations create a dependence relationship between donors and recipients, reinforce structural constraints to development, perpetuate ethnic strife and armed conflicts, and function as a poor and episodic substitute for long-term investments.[20] Moreover, it has been argued that the current form of humanitarian relief, undertaken almost exclusively by nongovernmental organizations, relieves rich governments from the need to search and identify the root causes for catastrophes and from the duty to mobilize the resources that may be generated to tackle such causes. In the words of one observer, "It is not rare for a refugee to get more food aid and protection than the ('simply starving') population in surrounding villages."[21] Not unlike concurrent developments in the field of CSR, efforts to create new legal norms for humanitarian intervention, and efforts to transcend the relief–development divide by means of novel forms of global regulation, are so far unsuccessful. First, principles of sovereignty still dominate international law, and the willingness to compromise them is slow to emerge. Second, rich states are extremely reluctant to commit human and material resources for universal causes disengaged from their concrete strategic interests. Under such circumstances, the "right" to deploy armed humanitarian intervention remains, at this point at least, the prerogative of powerful states that use (or some would say abuse) the humanitarian cause on a highly selective basis. As the latest U.S. led war in Iraq so clearly showed, humanitarian intervention can easily become a trope for violating international law in the name of a higher moral duty rather than an opportunity for validating a truly global regime of human rights.

On Humanitarianism and Corporate Social Responsibility

It is through their assigned locations within the neoliberal conceptual framework of the relationship among states, markets, and civil society that we can appreciate the striking similarly between prevalent practices of humanitarianism and contemporary corporate-inspired notions of social responsibility. Both humanitarianism and corporate social responsibility are technologies of intervention that have been constituted around principles of "metacharity." Both these technologies of intervention tend to operate "outside the law" and to emphasize voluntary action and goodwill alone. Both, albeit with different histories and orientations, are children of the neoliberal version of civil society. The emergence of humanitarianism as a major venue of disaster relief strongly corresponds to the neoliberal ideology of constituting a "free society" on the pillars of the voluntary and spontaneous action of responsible individuals. From this perspective, we can appreciate the typical reliance of humanitarianism on voluntary nongovernmental organizations and on the typical delivery of relief on a no-fault basis that strictly adheres to principles of neutrality. Corporate social responsibility, on its part, is a direct consequence of the shifting relations between markets and states. Multinational corporations, widely conceived as the primary beneficiaries of the neoliberal global order, are increasingly expected to assume social tasks that heretofore fell on the shoulders of national governments. The emergent new discourse of corporate social responsibility addresses the moral duty of corporations to act as "good citizens," to adhere to labor and environmental standards, and to contribute to the development and well-being of local communities and indigenous populations according to the neoliberal agenda of relieving governments from a host of similar responsibilities. That is, the spread of neoliberalism as a dominant blueprint of governance expresses itself in voluntary displays of corporate responsibility "in civil society" that function as a substitute for formal laws and regulations.

Yet as mentioned earlier, both CSR and humanitarianism are also sites of concrete and conceptual struggles. Corporate social responsibility is envisioned by some as an opening for a future global regime of enforceable rules shaping corporate governance and as a pretext for the emergence of legally protected norms of social responsibility. At least some confrontational nongovernmental organizations and activist groups try to couple corporate social responsibility with legal responsibility and to translate corporate responsibility into the language of blaming and claiming. Humanitarianism, on its part, is also undergoing some serious rethinking, as new calls for its politicization, legalization, and release from its charitable relief entrapment are increasingly heard.

Having established the structural homology between humanitarianism and CSR as two types of metacharity operating within the neoliberal framework, and having outlined some ideas about the need to relocate both CSR and humanitarianism within the formal domain of law, we can now consider the possible interface between the two. Therefore, we may ask: How may we conceive the actual and potential obligatory role of MNCs in alleviating or preventing distant suffering? In spite of the principled correspondence between humanitarianism and CSR as charity-based forms of social giving, and in spite of the growing recognition that MNCs play an increasingly significant political and social role in world events, little attention has been given to this question. To date, most discussions of humanitarianism relate to states and member-state organizations on the one hand and to nongovernmental organizations on the other hand. Most discussions of CSR almost consistently avoid the question of catastrophes. To some extent, this lack of discussion may simply reflect an assumption that the role of MNCs in catastrophes is in perfect accord with their routine charity-based campaigns. From this perspective, we may not expect MNCs to go beyond contributing to emergency relief efforts on a sporadic basis. However, this lack of discussion may also be revealing in terms of the distinction between relief and development. Conceived primarily as economic entities, corporations are associated with activities relevant for development projects much more than for emergency relief operations. As we shall see in the next section, the persistence of this conceptual divide paradoxically relieves corporations from any special humanitarian input save for the aforementioned displays of goodwill donations.

The Business Humanitarian Forum: A New Market-Society Blueprint?

The Business Humanitarian Forum (BHF) represents a novel attempt to introduce MNCs to humanitarianism. Launched in Geneva in January 1999, BHF's purpose is "to encourage dialogue and mutual support between the business and humanitarian communities, based on their common interests in the stability, prosperity and democratic evolution of developing societies and countries in transition." Acknowledging that only limited channels of communication exist between business and humanitarian organizations, BHF was founded by a group of corporate executives (e.g., Merck, Pfizer, Nestlé, Shell, Mobile, and Unocal) and humanitarian actors (e.g., Care USA, ICRC, and some former officials of UN relief agencies). BHF also established a formal partnership with the United Nations Development Program (UNDP).

BHF's model of cooperation merits some consideration. The model consists of four stages, each containing explicit and implicit assumptions about the very meaning of catastrophes, about victims of catastrophes, about the nature of humanitarianism, and about corporate responsibility. The model's point of departure treats a catastrophic event as a "disturbance" that affects, in this order, humanitarian organizations, business groups, and "people in the region." In the first stage, the catastrophe interrupts the ongoing development of a society because the destroyed infrastructure also means that "normal business operations and investment are impeded or completely stopped." The second stage exists immediately after the crisis strikes, when "humanitarian organizations intervene to help the affected community cope with the situation through emergency assistance involving food, medicine, shelter and other supplies." The third stage is intermediary. It describes an assumed phase when emergency relief exhausts itself and humanitarian organizations "seek to stabilize the society and restart the development process. Business becomes interested in resuming normal activity but hesitates to do so." It is not exactly clear from the model what brings this phase to an end. A fourth stage is launched, however, when "the affected community recovers enough to resume normal social and economic development." At this point, humanitarian organizations begin to leave and are supposedly replaced by the development programs of various bilateral and multilateral aid agencies. At this stage, according to the model, "multinational corporations can assist the local community through resumption of investment and business development."[22]

It is noteworthy that the victim population has almost no role in this model. Recovery is a function of passive acceptance of relief, and resumption of "normality" results from external factors. The model also strictly observes the relief–development divide in terms of both social organization and temporality. BHF states that as "the role and influence of governments continue to decline, business is becoming an actor in new fields." Accordingly, and to facilitate the needs of humanitarian organizations, MNCs may offer money and contribute their "pragmatic thinking, expertise and technology transfer, as well as practical assistance during crises when business can respond with less red tape [than governments]." The model thus explicitly ties MNCs to relief efforts through the idea of corporate social responsibility in its charity-oriented version, stating that such assistance will serve as an updated display of responsibility because "corporate citizenship has evolved over the last two decades from occasional philanthropic contributions to a more systematic and focused approach." The CSR model that is here tied to humanitarianism is also explicitly utilitarian, applying an instrumental rationality to humanitarian intervention. BHF states that "multinational corporations can achieve a strategic business benefit through humanitarian activities This is part of the rationale behind corporate social responsibility or corporate citizenship." Linking CSR to humanitarianism, according to BHF, "can provide clear business advantages": Above and beyond their direct investments that are endangered by a catastrophe, MNCs depend in part on the local workforce, local businesses, and host country leaders. Therefore, "multinational corporations can have very direct incentives for supporting the relief efforts of humanitarian organizations." Further, BHF states that displays of corporate citizenship in the form of helping humanitarian organizations are rational because MNCs rely on them to "stabilize societies and make long-term investments possible."

Situating humanitarian organizations squarely on the side of "relief" and offering charity-based corporate assistance to such relief efforts, the BHF model also squarely locates MNCs on the side of "development" in its most traditional sense—namely, investment, generation of revenues, and, at best, facilitating new jobs and economic opportunities. The model assumes that renewed corporate activity in and of itself launches a process of development and that this process, in turn, enhances the self-reliance of the population. However, we now know that development approaches often overlook long-term sustainability concerns and are driven by a search for quick profits. Thus, development projects rely on the stronger elements in society, often bypassing the poorest and most needy. Moreover, development projects are often undertaken by foreign experts and involve little effort to train the local workforce and engage in technology transfer. Indeed, when we take a closer look at the model that BHF offers, we see that once development by MNCs is considered, the operational rationale clearly changes from that of business responsibility to that of business opportunity.

The partnership model of BHF also assumes a temporal linearity according to which MNCs suspend commercial activities when a catastrophe strikes, offer charity-based support for humanitarian organizations until the society is "stabilized," and only resume operations after significant recovery has already been achieved. The problem with this model, however, is that a catastrophe is rarely a single event at a single moment in time. Rather than following clear progressive stages, a catastrophe is a complex process in the course of which the lines between relief and development are constantly blurred, and progression and regression are interchangeable. For example, it is now widely acknowledged that famines are not the result of natural causes alone. The natural–social divide in general is seriously challenged by latter-day students of disasters.[23] The dissolution of this distinction is crucial because when social causes contribute to catastrophes, there are more reasons to actively protect and enforce victims' human rights. Accordingly, famines are now treated as a prototypical example of a complex humanitarian crisis. Food shortages and crop failures, the immediate causes for widespread hunger, are often caused by war, mass deportations and forced displacements, politics of food distribution, unavailability of foreign relief supplies, corruption, failed infrastructures, and human-made desertification. Development projects in which foreign entities collaborate with corrupt or abusive regimes also contribute to the complexities of famine as they sometimes trigger or enable forced displacements, the destruction of agriculture, and the abuse of material resources for furthering hostile political agendas. Humanitarian efforts often reveal these complexities. Whether humanitarian aid is delivered through international bodies like the United Nations, specific national governments, or nongovernmental agencies, the availability of food and its methods of allocation depend on the local government, on its conception of the degree to which humanitarian intervention infringes upon sovereignty, on the ability of the interventionist bodies to enjoy a supportive political infrastructure, and on the relationship between ruling elites and MNCs that operate in the country.

Finally, BHF mentions that the involvement of MNCs in some countries' local politics "had a serious negative impact on the local humanitarian situation." But this passing reference does not lead to any substantive discussion about the potential contribution of MNCs to catastrophes and about the potential ability of MNCs to prevent catastrophes before they occur. Now, at least since the catastrophic results of Union Carbide's operations in Bhopal, it may amount to a truism to discuss the potential direct responsibility of corporations to mass disasters.

In what follows, therefore, I do not discuss such risky potentials but rather focus on situations in which corporate responsibilities for catastrophes or for their prevention are indirect and silent. To illustrate, let us turn back to Talisman Energy Inc., whose social responsibility report was briefly introduced in the beginning of this chapter, and discuss some of its activities and their consequences in Sudan.

Oil, Famine, and Law

A civil war has persisted in Sudan since the 1980s. Rebel armies in southern Sudan have been fighting Sudanese government forces and progovernment militias in a bid for political autonomy for Sudan's primarily Christian population of 5 million. Violent ethnic and military divisions among southerners have further complicated the civil war and worsened the plight of the local population. Over the years, and coupled with serious droughts, the civil war has created mass population displacements, leading to serious food shortages and widespread famines in 1988, 1992, and 1998. Throughout, food has been used as a weapon by both government and rebel forces. Various independent studies concluded that, over the years, government officials have placed tight controls on aid deliveries, often blocking food shipments to needy populations, while many rebel commanders regularly have confiscated a percentage of food relief distributed in the south. It was also established that the Sudanese government had sold grain reserves to fuel their military, while refusing to declare a food emergency and to allow relief into starving opposition areas. Both government and opposition forces created famine as a tool to control territories and populations and restricted access to food aid (often by attacking relief convoys) as an instrument of ethnic and religious oppression.[24]

One of the conflict areas is southern Sudan's Upper Nile province. Mass displacements, gross violations of human rights, and ensuing hunger in the Upper Nile province were directly related to clashes around the region's vast and largely unexploited oil fields. The oil fields in Western Upper Nile are crucial to the government's ability to generate revenues. In 1998, construction was completed on the pipeline to carry the crude oil to refineries in the north, provoking the resistance of southerners who saw this oil as their property and who considered the government's reliance on oil revenues a prime factor enabling it to persist in the war. Ever since, as several reports indicate, the government has pursued a "scorched earth" policy to clear the land of civilians and to make way for the exploration and exploitation of oil by foreign oil companies. Subsequently, it is claimed that the Sudanese government's revenues from oil have enabled it to double its military expenditures in 2001 compared to 1998. Another report thus claims:

> That this fabulous potential for oil wealth exists side by side with a famine that affects more than 150,000 people in Western Upper Nile is no accident. It is the consequence of government desire to establish control over the area by using militias (since 1983) to loot and attack and displace the local population. The 1998 Western Upper Nile famine has been largely the product of unrestrained attacks on the civilian population by two pro-government militias.[25]

Sudan's oil production and sales have been undertaken by a consortium of multinational oil companies that was put in charge of the $1.6 billion oil development scheme. In 1998, Talisman became a major player in the consortium. Subsequently, Talisman reported that the Sudan operations generated a pretax income of Can$310 million in 2002, up from Can$210 million in 2001. A 2001 report by Corporate Watch claimed that the participation of Talisman in the consortium was particularly significant not only because it provided the technical expertise needed to build a new 900-mile pipeline to Port Sudan on the Red Sea but also because it carried the stature of a Western oil firm.

There seems to be no dispute over the fact that as late as 2002, the Upper Nile region had been a hunger zone. Reports from UN agencies, the World Food Program, and the U.S. Committee for Refugees all shared grave concerns about the humanitarian disaster in the region. What seems to be disputed is the complicity of the oil companies in facilitating the ongoing catastrophe. As mentioned, one common argument is that the revenues from the development of the oil fields and oil pipelines prolong the war and increase the likelihood of recurring famines. It is also argued that the Sudanese government divided the south into a web of oil concessions, provoking the turning of each bloc into a potential battlefield because the development projects in the oil fields are premised on mass displacements that both fuel hostilities and create acute food shortages. Even more concretely, critics argued that the oil companies operating in Sudan were complicit in displacements and in other human rights violations due to their involvement in the government's operations. It has been argued that the companies allowed the government forces that were assigned to protect them to use their airstrips and roads to displace people and to commit a variety of war crimes.

Thus, although Sudan has largely been perceived as a chronic disaster area that merited humanitarian assistance, the focus on oil production activities profoundly influenced the way the Sudanese catastrophe was perceived. The shifting of the gaze extricated the issue from the encompassing universe of humanitarian relief and moved it into the universe of rebellious humanitarianism: a technology of intervention that—while constantly professionalizing the alleviation of suffering—seeks to uncover the political and economic roots of catastrophes and to mobilize the resources needed to remove them. Within this context, attention is drawn to the role of multinational corporations in creating, directly or indirectly, conditions of suffering. The emerging discourse and practice of corporate social responsibility—based on the widespread understanding that some MNCs currently enjoy policy-shaping powers that match and often surpass that of national governments—further facilitate this conceptual shifting of the gaze. The focus on Talisman is therefore interesting because, whether it is the charity-oriented nature of humanitarianism or the charity-oriented nature of CSR that is challenged, this focus is at the same time a call for bringing both into the domain of legal obligations. In the Talisman case, humanitarianism and CSR converge through the language of human rights and the universal obligation to protect them. The Talisman case, in this respect, allows us to begin asking questions about the type and scope of duties that corporations may be expected to follow in catastrophic situations. Further, it allows us to ask what type of legal tools should be available when corporations fail to adhere to such duties.

Not surprisingly, the oil companies in general and Talisman in particular have fiercely denied allegations of wrongdoing while trying to preserve the line between charitable actions and legal obligations. Talisman argued that the oil fields it operated had been vacant of civilian populations before it entered the area and that, rather than exacerbating the conflict, it had made a considerable contribution to peace efforts. Launching a vigorous corporate responsibility campaign, Talisman pointed to the modern infrastructure it built for the local population: new water wells, schools, clinics, and a well-equipped hospital that included an operating room and a neonatal unit. As mentioned in the opening section of this chapter, Talisman reported a host of community programs taking place in Sudan, small business initiatives, educational programs, and active involvement in peace efforts initiatives. The president and CEO of Talisman had declared in October 2002 that

> Talisman's presence in Sudan has been a force for good and we have taken steps to ensure that the benefits created through our involvement will continue to improve the lives of the people of Sudan both now and in the future. Talisman and its employees have made significant contributions to this end over these past four years, providing medical assistance, shelter, clean water, vocational training and initiating capacity-building programs.[26]

Finally, Talisman has repeatedly declared that its policies in Sudan promoted and protected human rights because they were based on the principles of the Universal Declaration of Human Rights.

Unconvinced critics included a number of human rights, religious, and humanitarian organizations. From a theoretical point of view, these critics fused humanitarian concerns with expectations of corporate social responsibility, invoking the language of human rights as a political and legal medium. In the case of Sudan, according to this logic, corporate social responsibility amounted to no less than complete operational withdrawal as a means for alleviating or at least preventing further catastrophic results. In practical terms, this fusion relied on an effort to bring both CSR and humanitarianism into the domain of law.

Thus, apart from a general moral shaming campaign, critics mobilized economic and legal pressure aimed at forcing Talisman to withdraw from Sudan. Economic pressure was in the form of a vast divestment campaign, mainly addressing Canadian and U.S. shareholders. The divestment campaign was initially aimed at the Ontario Teachers' Pension Plan, Talisman's largest investor. Accordingly, the executive board of the Ontario Teachers Federation—the umbrella organization for Ontario's 144,000 teachers—has asked the Ontario Teachers' Pension Fund to sell its holdings in the company. In early 2000, the state of New Jersey Finance Board announced that it had sold all its 680,000 shares in Talisman. The General Assembly Council of the Presbyterian Church (USA) approved a recommendation from the National Ministries Division Committee to add Talisman to the denomination's divestment list, barring church entities from owning stock in the company. By mid-2000, one report noted that several important investors had already pulled out, including the state of New Jersey, the New York City Pension Fund, the California Public Employees' Retirement Fund, the Texas Teachers' Retirement Fund, TIAA-CREEF Investment, and Investors Group Investment Management.[27] And in 2001, the pension committee of the Anglican Church of Canada also began to divest its holdings in Talisman.

On the legal front, Talisman had been challenged and warned by the International Centre for Human Rights and Democratic Development (Rights & Democracy), a Canadian nonprofit organization. At a news conference in 2002, Rights & Democracy warned Talisman that future complicity in Sudanese human rights abuses could make it liable for prosecution by the newly established International Criminal Court. Rights & Democracy further argued that although the Rome Statute of the International Criminal Court excluded "legal persons" (e.g., corporations) from being tried under its jurisdiction, corporate executives who facilitated, aided, or abetted a crime covered by the court might be criminally liable. It argued that after entering into effect in July 2002, crimes committed by state agents or nationals of states that have ratified the statute will be liable for prosecution. Canada, it reminded Talisman, was one of the sixty-six countries to have ratified the Rome Statute and incorporated it into Canadian law, which means that such suspects could also be tried in Canadian courts.[28]

Even more significantly, in November 2001, the Presbyterian Church of Sudan, aided by lawyers and activists in Canada and the United States, had filed a class-action suit against Talisman in the U.S. District Court for the Southern District of New York.[29] Plaintiffs argued that defendants have collaborated with the Sudanese government in a joint strategy to deploy military forces in a brutal ethnic cleansing campaign against the civilian population "for the purpose of enhancing defendants ability to explore and extract oil from areas of southern Sudan by creating a *cordon sanitaire* surrounding the oil concessions located there." Plaintiffs further argued that the armed campaign was made possible through government utilization of vehicles, helicopters, aircraft, roads, and airstrips owned, chartered, constructed, or maintained by Talisman. Plaintiffs accused Talisman of keeping a blind eye to military operations that relied on Talisman's resources and were done in the name of protecting the oil fields. These military operations resulted in severe violations of human rights and obligatory norms of customary international law, including killings, rape and torture amounting to genocide, ethnic cleansing and the displacement of more than 100,000 people, vast destruction of property and crops, and the deliberate obstruction of health and food distribution undertaken by humanitarian organizations. Plaintiffs asked the court to declare that Talisman violated international human rights law, to issue an injunction restraining it from continuing cooperation with the Sudanese government, and to establish compensatory and punitive damages.[30] The claim against Talisman is still pending, but it should be noted that there is mounting pressure in the United States to prevent the use of national legislation for suing MNCs, and the future of using American courts as a means for holding MNCs accountable for human rights violations remains highly uncertain.

Responding to the legal and nonlegal pressure to which it had been subjected, Talisman announced in early 2003 that it sold its interests in the Greater Nile Oil Project to an Indian corporation for approximately U.S.$771 million. In the next and final section of this chapter, I briefly discuss the potential meaning of—and the future prospects of—invoking the law as a means for tackling the role of corporations in complex humanitarian disasters of the type discussed here.

Conclusion

International law treats complex humanitarian catastrophes largely through a human rights perspective. It is the violation of human rights that is currently invoked to justify humanitarian intervention and to identify perpetrators who violate them. Hunger itself is considered a human rights issue in humanitarian covenants that provide for feeding civil populations during wars. International legal instruments, such as the International Conference on Nutrition World Declaration and Plan of Action for Nutrition (1992) and the Vienna Declaration on Human Rights (1993), support the principle that food should never be used as a political tool and that hunger should never serve as a weapon. They provide a reference point and standard for action for the United Nations, its member states, and its agencies. The second, third, and fourth Geneva Conventions (1949) and Additional Protocols (1977) also provide international guidelines to combatant parties for meeting essential humanitarian needs and ensuring basic subsistence rights of civilian populations experiencing armed conflict. All these legal norms follow the human rights principles expressed in the UN Charter and Universal Declaration of Human Rights that treat food as a basic human right and as a principal component of the universal human right to life.

The international law of human rights is therefore also the primary venue in the effort to transform the heretofore voluntary and nonbinding obligations of MNCs in the general field of CSR and in the specific area of humanitarian catastrophes into enforceable forms of legal accountability. However, international law mainly addresses the duties of governments, and MNCs have so far remained outside contemporary debates. Only in recent years, to some degree, have prominent jurists taken up the task of trying to create a new architecture of domestic and international law addressing the human rights obligations of multinational corporations.[31] Thus, the legal action against Talisman must be understood as one element in a wide spectrum of attempts to tame corporate behavior by inventing new global regulatory regimes that would also address the potential obligations of MNCs to alleviate suffering and to assume active humanitarian duties.[32] The resistance of corporations to the use of binding legal obligations, on the other hand, must be correspondingly understood as an element in corporate attempts to locate CSR (including humanitarian responses to catastrophes) within the voluntary domain. Indeed, throughout this chapter, I tried to show that both technologies of intervention—CSR and humanitarianism—neatly correspond with the ascendance of neoliberalism. Both nurture an ideal civil society that assumes voluntary and spontaneous actions for correcting social wrongs, while relieving governments and markets from assuming legally binding duties. Both CSR and the ideology of the new humanitarianism, as Chimni puts it, seek "to legitimize and sustain an international system that tolerates an unbelievable divide not only between the North and the South but also inside them." [33] Out of this conjectural relationship between CSR and humanitarianism, an inquiry as to the role of MNCs in catastrophes must be rethought. As we have seen, new demands for legalizing and for establishing a right to humanitarian intervention have been firmly placed on the international agenda in recent years. Still, most attention is given to the actions of states. Yet isn't it time to begin to think, even prior to armed humanitarian intervention, about creating a legal framework for sanctioning corporate humanitarian intervention? In fact, although intervention may be a humanitarian instrument for governments, the act of withdrawal—the practice of "mis-intervention"—may become the appropriate humanitarian instrument when it comes to corporations. Often well ahead of foreign countries in terms of access, resources, and connections with relevant agents in disaster-stricken areas, MNCs are in an increasingly strategic position to anticipate and prevent catastrophes and to contribute to their alleviation in many host countries. The duties of MNCs that operate in such areas must therefore be urgently addressed. Thus, by speaking about corporate humanitarian intervention and withdrawal, we may begin to imagine the articulation of active corporate duties to intervene in the course of catastrophic events (1) by barring host governments from using corporate facilities and know-how, (2) by requiring host governments to enter into enforceable security protocols that may be independently monitored as a condition for investment and development projects, and, when necessary, (3) by pulling out and suspending operations in areas where they exacerbate catastrophic conflicts. Current debates about humanitarian intervention still almost entirely avoid these issues. As we have seen, this may be due to the fact that when it comes to corporations, both CSR and humanitarianism are still by and large conceived, projected, and advocated as the nonbinding affairs that civil society may and should promote without the assistance of formal law.

Notes

1. Ronen Shamir, "The De-Radicalization of Corporate Social Responsibility," *Critical Sociology 30*, no. 3 (2004), 669; Ronen Shamir, "Mind the Gap: The Commodification of

Corporate Social Responsibility," *Symbolic Interaction 28*, no. 2 (2005), 229; Ronen Shamir, "Corporate Social Responsibility: A Case of Hegemony and Counter-Hegemony," in *Law and Globalization from Below: Towards a Cosmopolitan Legality*, Boaventura de Sousa Santos and César A. Rodríguez, eds. (Cambridge: Cambridge University Press, 2005); David Vogel, *The Market for Virtue: The Potential and Limits of Corporate Social Responsibility* (New York: Brookings Institute Press, 2005).

2. Morton Winston, "NGO Strategies for Promoting Corporate Social Responsibility," *Ethics and International Affairs 16*, no. 1 (2002), 71.

3. Pall A. Davidsson, "Legal Enforcement of Corporate Social Responsibility Within the EU," *Columbia Journal of European Law 8* (Summer 2002), 529.

4. Andrew Clapham, "The Question of Jurisdiction Under International Criminal Law over Legal Persons: Lessons from the Rome Conference on an International Criminal Court," in *Liability of Multinational Corporations Under International Law*, Menno T. Kamminga and Saman Zia-Zarifi, eds. (The Hague: Kluwer Law International, 2000).

5. Sara Joseph, *Corporations and Transnational Human Rights Litigation* (Oxford: Hart Publishing, 2004); Ronen Shamir, "Between Self-Regulation and the Alien Tort Claims Act: On the Contested Concept of Corporate Social Responsibility," *Law and Society Review 38*, no. 4 (2004), 635.

6. Merck press release, March 7, 2001, www.merck.com.

7. Associated Press, March 4, 2000, www.FreeRepublic.com.

8. Ronen Shamir, "Corporate Responsibility and the South African Drug Wars: Outline of a New Frontier for Cause Lawyers," in *The World Cause Lawyers Make: Structure and Agency in Legal Practice*, Austin Sarat and Stuart Scheingold, eds. (Stanford, CA: Stanford University Press, 2004).

9. Adi Ophir, "Moral Technologies: Managing Disasters and Forsaking Life," *Theory and Criticism 22* (2003), 67 (Hebrew).

10. Luc Boltanski, *Distant Suffering: Morality, Media and Politics* (Cambridge: Cambridge University Press, 1999), 191.

11. Jerry Buckland, "From Relief and Development to Assisted Self-Reliance: Nongovernmental Organizations in Bangladesh," *Journal of Humanitarian Assistance* (1999), http://www.jha.ac/articles/a052.htm.

12. Pierre Senarclens, "Neo Liberalism and Humanitarianism," unpublished paper presented at the Law and Catastrophe Workshop, Tel Aviv University (2003), 5.

13. Lucy Taylor, "Globalization and Civil Society—Continuities, Ambiguities and Realities in Latin America," *Indiana Journal of Global Legal Issues 7*, no. 1 (Autumn 1999), 269.

14. Mary Kaldor, "Transnational Civil Society," in *Human Rights in Global Politics*, Timothy Dunne and Nicholas J. Wheeler, eds. (Cambridge: Cambridge University Press, 1999).

15. Boltanski, *Distant Suffering*, 178.

16. Michael Walzer, "Beyond Humanitarian Intervention: Human Rights in Global So-

ciety," The Minerva Center for Human Rights, Annual Lecture Series on Human Rights, Tel Aviv University School of Law, June 16, 2004.

17. B. S. Chimni, "Globalization, Humanitarianism and the Erosion of Refugee Protection," *Journal of Refugee Studies* 13, no. 3 (2001), 243.

18. Mahmood Mamdani, "Humanitarian Intervention: A Forum," *The Nation*, July 14, 2003.

19. Boltanski, *Distant Suffering*, 191.

20. Mark Duffield, "The Political Economy of Internal War: Asset Transfer, Complex Emergencies, and International Aid," in *War and Hunger: Rethinking International Responses to Complex Emergencies*, Joanna Macrae and Anthony Zwi, eds. (London: Zed Books, 2004).

21. Senarclens, "Neo Liberalism and Humanitarianism."

22. http://www.bhforum.ch/en/partnership/brochure_03.cfm.

23. Amartya Sen and Jacques H. Dreze, *The Political Economy of Hunger* (Oxford: Clarendon Press, 1995).

24. David Keen, *The Benefits of Famine: A Political Economy of Famine and Relief in Southwestern Sudan, 1983–1989* (Princeton, NJ: Princeton University Press, 1994).

25. *The Scorched Earth: Oil and War in Sudan*, Christian Aid Report, http://www .christian-aid.org.uk/indepth/0103suda/sudanoil.htm.

26. Talisman press release, "Talisman to Sell Sudan Assets for C1.2 Billion," Calgary, October 30, 2002, http://www.highbeam.com/doc/1G1-132318696.html.

27. Wayne Sawtell, "Divestment Campaign Targets Talisman," *Peace and Environment News*, April 2000, http://perc.ca/PEN/2000-04/sawtell.html.

28. Andrew Clapham, "The Question of Jurisdiction Under International Criminal Law over Legal Persons: Lessons from the Rome Conference on an International Criminal Court," in *Liability of Multinational Corporations Under International Law*, Menno T. Kamminga and Saman Zia-Zarifi, eds. (The Hague: Kluwer Law International, 2000). Also www.corpwatch.org/bulletins/PBD.jsp?articleid=2478; http://www.ichrdd.ca.

29. Civil Action No. 01 CV 9882 [DLC]. See amended complaint at http://www .bergermontague.com/pdfs/SecondAmendedClassActionComplaint.pdf.

30. The class-action suit was submitted under the Alien Tort Claims Act (Judiciary Act of 1789, Ch. 20, §9, 1 Stat. 73, 77, 1789, currently with minor changes, 28 U.S.C. §1350, 1982).

31. Steven R. Ratner, "Corporations and Human Rights: A Theory of Legal Responsibility," *Yale Law Journal* 111, no. 3 (2001), 443; David Kinley and Junko Tadaki, "From Talk to Walk: The Emergence of Human Rights Responsibilities for Corporations at International Law," *Virginia Journal of International Law* 44, no. 4 (2004), 931.

32. Ronnie Lipschutz and James K. Rowe, *Globalization, Governmentality, and Global Politics: Regulation for the Rest of Us?* (New York: Routledge, 2005).

33. B. S. Chimni, "Globalization, Humanitarianism and the Erosion of Refugee Protection," *Journal of Refugee Studies* 13, no. 3 (2001), 245.

Political Catastrophe and Liberal Legal Desire: Two Stories of Revolution, Remediation, and Return from the French Nineteenth Century

SYLVIA SCHAFER

What counts as catastrophe? What kind of difference does it denote? What kind of desire has that difference awakened in relation to law? In many respects, catastrophe's range of meaning in the West has remained relatively stable since at least the late eighteenth century. In the modern era, "catastrophe" tends to be used most often to describe profoundly destructive and unexpected disasters, occasions of calamitous damage, disruption, or loss. Those who invoke or explore the sign of catastrophe have been especially concerned with the discursive separation of catastrophe from more "ordinary" devastation. In the wake of the 1995 earthquake in Kobe, for example, the World Health Organization emphasized scale in its discussion of the catastrophic. "Not all earthquakes are endowed with the dimensions of a catastrophe," the organization's press release on a 1997 meeting of experts explained. Instead, catastrophe should describe only those occasions "where the loss of human life, the material damage and the destruction of health and other services are such that extraordinary aid must come from outside the affected area." [1] Working along similar lines, the humanitarian group Doctors Without Borders has used "catastrophe" to describe "a sudden and relatively brief event, affecting a collectivity, injurious to people and their property." [2] Natural catastrophe, the organization's manual on "emergency care in catastrophic situations" explains in more detail, "brutally interrupts the course of daily existence. Reduced to powerlessness, plunged into distress . . . the population is left without provisions, clothing, shelter or sanitation." [3] The manual affixes the term *catastrophe* to those human events that result in similar devastation, such as armed conflict. [4] Catastrophe is also read in its symptoms: Human victims of catastrophe, whether the event is of natural or other origin, often display exaggerated symptoms of "disaster syndrome," including "inhibition, stupefaction, weeping, hysteria and depression." [5]

The precise delimitation of catastrophe's difference also emerges as a pre-occupation of the modern "insurantial imaginary."[6] As one industry manual puts it: "[a] catastrophic loss" is best defined as "one that is a sudden, widespread, or extraordinary disaster. The two essential characteristics of a catastrophe are that it be (1) sudden and (2) widespread."[7] Like Doctors Without Borders, this text's authors contend that the "origin of catastrophes may be either human or natural. The human causes are war; invasion; civil war; insurrection; rebellion; revolution; military, naval, or usurped power; civil commotion; conspiracy, riots; strikes; and martial law. The natural causes are movements of earth, air, or water."[8] For twentieth-century insurers, catastrophe names a class of risk beyond the bounds of the ordinary policy, although not beyond the limits of insurability.[9] Thus, a recent French encyclopedia of insurance states that "catastrophic risks (risks of war or uprising; natural catastrophes; nuclear risks . . .) are in principle excluded from classic insurance coverage in all European countries."[10] Subsequent sections review catastrophe risk as it has been renegotiated in the second-tier reinsurance market.[11]

Across these modern renderings, four central characteristics tend to designate catastrophe's difference. First, catastrophe demands narratives of ordinary life suddenly and brutally disrupted by the drastic event.[12] It is this use that most closely draws on the word's Greek roots, *kata-* and *strophé*, which translate literally as "down- (or over-)turning." It also descends from early modern notions of catastrophe as the plot reversal that brings a stage drama to its end, especially but not uniquely in tragedy.[13] Suddenness, however, also frequently merges with its opposite. The abruptness supposedly essential to naming the catastrophic incident often disappears into the account of those lasting and widespread effects that define catastrophe retrospectively. The temporal moment of rupture, as the group Doctors Without Borders suggests, stretches into an apparently endless time of privation, "stupefaction," "powerlessness," and "distress." Narratives of catastrophe collapse the instant and the eternity. They extend the event toward the horizon of the future and read it as an origin in anticipation of the time that will be after the fact.[14]

Modern discussions of catastrophe commonly presume an equivalence between catastrophes with "human origins," such as war and revolution, and those associated with the violence of the natural world.[15] In this regard, the imaginary space beyond that of the "ordinary" disaster is brought into view more through reflection on the analogies of devastation than on some intrinsic qualities of events in themselves. This openness to interchangeability, the subsidence of one set of tropes into another, speaks as well in the movement of figures of speech between the domains of natural and human catastrophe where observers struggle with the problem of representing catastrophe's difference. The figurative oscillation between the human and the natural is also increasingly evident where the explanatory power of the divine is shaken or displaced as a primary point of reference in accounts of earthly devastation.[16] The French revolution of 1848, a political and social "catastrophe" in the eyes of Alexis de Tocqueville, required the language of natural violence. The unresponsive King Louis Philippe, Tocqueville wrote, was no more than "a man awakened at night by an earthquake, who, seeing his house falling down in the darkness and even the ground giving way under his feet, remains distracted and lost amid the universal unforeseen ruin."[17] In his stress on the "unforeseen," Tocqueville employed the natural analogy to elaborate a temporal dimension for political catastrophe: Like an earthquake that strikes without warning, the sudden intrusion of the catastrophic event rudely destroys an imagined future that gives meaning to the present as its secure horizon.[18]

Whether the narrative of sudden change and its aftermath is analogized to nature or not, catastrophe often serves to signify a disturbance marked by overwhelming excess.[19] This excess also challenges the limits of representation.[20] It is that which demands "extraordinary" response. It emerges in relation to risk or damage that flows beyond the limits of the normal or the acceptable.[21] Its scale and scope spill beyond expected boundaries and demand response from afar, dissolving the meaning of geopolitical borders as much in the production of horror as in the mobilization of intervention or aid. The quality of excess emerges also in those efforts to enumerate catastrophe that end with the signs of the unspoken or unspeakable, the ellipsis of a future yet to be written: "risks of war or uprising; natural catastrophes; nuclear risks . . . [sic]"[22] "A twig bends until it breaks, a seed withers or germinates. An unstable compound explodes. The earth quakes. And so on."[23]

At the same time that it signals crises of excess, the category of catastrophe also often accompanies stories of unimaginable deprivation. This is true not only with regard to basic sustenance and the supply of material goods but also, as Doctors Without Borders suggests in its invocation of "inhibition, stupefaction, weeping, hysteria and depression," a profound loss of self. Tocqueville's account of his own disorientation, of the sudden remaking of identities and relations, articulates a sense of catastrophe in which many more than the king are left "distracted and lost amid the universal unforeseen ruin." Although a great deal of scholarship in psychology, literary studies, and history has addressed the question of self and catastrophe through the analytical framework of memory and trauma, Tocqueville's emphasis on the political underscores the importance of considering catastrophic devastation to selves conceived and narrated along other discursive axes. [24] One of the most historically powerful alternative lines of imagining catastrophe and its consequences for the subject runs through the domain of law, and especially, as Tocqueville reminds us, across the ground of liberal reflection on political catastrophe, the constitution of coherent individual subjects, and the law's capacity to produce a moral, stable, and meaningful world. [25]

This chapter takes up the historicity of "law and catastrophe" by exploring the problem that political events, represented and narrated as human-made catastrophes, have posed for the liberal legal imagination during the turbulent first half of the nineteenth century in France. It focuses in particular on those political catastrophes that appeared to contemporary observers to result in an unbearable disintegration or dislocation of the legally meaningful civil individual: the catastrophes of revolutionary uprising and their long aftermaths. Thinking about law and catastrophe by reading liberal discourse and the desires it evinced in the nineteenth century offers an especially useful opportunity for examining a dominant and still potent strain of Western commitment to law as a means of securing the moral and political stability that in turn predicates progress, freedom, and individual self-determination. As the greatest challenge to that stability, the event named *catastrophe* has unearthed liberalism's fundamental desires, not only for freedom but also for order, authority, and the purity of boundaries and categories. [26] The discursive space of catastrophe likewise reveals where liberal visions of law have been activated in especially charged ways. Catastrophe is where liberalism has argued against its own shaken certainties for a renewal of faith in law's capacity to prevent, contain, and repair the effects of catastrophic political crisis. In this space as well, the historicity of those shifting arguments and that repeatedly shaken and renewed faith comes sharply into view.

Centering a discussion of catastrophe and law on historically situated visions of politics and the legal individual requires some careful thought about the analytical apposition of revolution and catastrophe. I do not bring these terms together by adopting a position of reaction or antirevolutionary repudiation, although this has of course been a common location for the discourse of revolution as morally and politically catastrophic. Rather, I write from an etymological and genealogical position that allows for more expansive reflection on stories of sudden redirections in political life and the long-lasting disturbances of identity that seem to ensue. [27] As in the case of catastrophe, readings of revolution and its effects entail the identification of drastic dislocations and reversals that upend a given order in unanticipated ways. "The modern notions of revolution," Hannah Arendt has famously argued, turns away from an older sense of regular, cyclical motion and instead recasts the "turning over" of revolution as an abrupt break with the past. By the end of the eighteenth century, the idea of revolution is "inextricably bound up with the notion that the course of history suddenly begins anew . . ." Revolution emplots "an entirely new story" whose outcome is either "victory or disaster."[28]

Revolution's representational and narrative potency as a catastrophe of human origin—as a sudden overturning that entails the collapsing of temporal, spatial, and figurative boundaries; the upending of given hierarchies; and the unbearable amalgam of loss and excess—deeply informs the work that liberalism has asked law to perform in its aftermath since the late eighteenth century, particularly in relation to the reconstruction and recalibration of the civil self. To explore the articulation of law's capacities and obligations at the beginning of this new era, I focus on two strongly charged discussions of how liberal constructions of law might mitigate political catastrophe and its long-lasting effects. Both date from the period between 1789 and 1850. The first, Honoré de Balzac's *Colonel Chabert*, considers the problem of civil existence after the revolutionary and Napoleonic breaks with France's monarchical past. The second, a set of juristic writings on the significance of state-provided legal aid around the revolution of 1848, addresses the catastrophic consequences of a poverty that excluded individuals from seeking rectifications through civil action and, once radicalized, threatened to drive them to the barricades. Punctuated by three regime-changing uprisings—the revolutions of 1789, 1830, and 1848—as well as by coups, large-scale protests, and periods of increasingly serious labor unrest, the first half of the French nineteenth century seemed to some observers to be in permanent confrontation with revolution. As Tocqueville put it in 1850, the episodes of revolution that had shaken France beginning in 1789 were in fact but one ongoing event, "always one and the same, through its various fortunes and passions, whose beginning our fathers saw and whose end we shall in all probability not see."[29]

This period is equally significant for its place in the genealogy of liberal desire for law in the face of catastrophe. Although the tools and framework for imagining responses to catastrophe in terms of risk management were rapidly being developed, questions of danger and damage were not yet fully integrated into the episteme of insurance.[30] Nor had the individual and his or her relationship to others in times of crisis been systematically woven into actuarial narratives organized by probability and the distribution of risk across space, time, and populations. The first half of the French nineteenth century thus occupies an important place in the "history of the present" by showing us a past for enduring liberal investments in law that was especially fraught by "catastrophic" challenges to law's moral and political stability but not yet shaped by the late nineteenth- and twentieth-century relocation of law alongside, and sometimes within, the techniques of probability calculations and rationalized indemnification.[31]

The past bears also on liberalism as a category of historical and political analysis. Readers may note that not every text considered here was generated by people who would have identified themselves as liberals. Balzac's novel, the most obvious instance, is valuable not because of any avowed authorial commitment to liberal principle but rather for the way it impales certain contemporary liberal desires for law on the spikes of its trenchant critique of those longings.[32] In the first section of the chapter, then, early nineteenth-century liberal hopes for law's restorative power in the aftermath of catastrophe are systematically, and sometimes cynically, refracted through Balzac's studious illiberalism. The second section turns to the voices of liberal thinkers and jurists more closely affiliated with what we have come to think of as liberal commitments to law, liberty, and the equality of individuals.[33] Despite France's recent history of repeated revolutionary upheaval, and even despite their own profound doubts, these men continued to voice the belief that law might yet contain the worst excesses of political catastrophe, exemplify and guide the progress of "civilization," and facilitate the realization of the individual subject within a well-ordered society.

Readers may note as well the moves across this chapter from fiction to memoir to jurisprudential treatise and parliamentary debate. Despite their apparent generic differences, these texts all serve as important points of entry into the rich legal imaginations of the period. More specifically, these works share the problem of how to narrate law's work in both the anticipation and wake of political catastrophe. They draw on the formal legal knowledge of their authors and rely on techniques of storytelling, scene setting, and creating figurative language as the means of discussing the political and moral import of law in a time of revolutionary upheaval. In this respect, the texts considered here all belong to the literature of law and catastrophe. From their diverse perspectives, and across the standard lines of genre or disciplinary object of inquiry, they all engage deeply with liberal concerns about law's capacity to repair the damage that abrupt political rupture and its aftermath inflicted on the legal individual, a figure celebrated by the revolution of 1789 and later installed by the Napoleonic civil code as a foundational element of the modern French juridical imagination.

As it reads explorations of revolution, law, and the civil individual in nineteenth-century France, this study underscores some of the historically specific particularities of French liberalism, especially its aversion to the leveling ambitions of republican and socialist democracy as they emerged on the scene of revolutionary upheaval, its unlikely affiliation—given its attachment to liberty and its association with "moderate" phases of revolutionary activism—with conservative visions of "order," and its commitment to French "civilization" as a moral alternative to the radical imaginary of the nation.[34] Despite their national inflections, however, the concerns that preoccupied liberal imaginations in nineteenth-century France have been influential if somewhat obscured elements in the larger domain of Western reflection on law and political catastrophes of the individual. In a broader view, then, this chapter explores the genealogy of both the perennially renewed liberal faith in the stabilizing effect of the legally constituted civil subject and the equally persistent anxieties about law's capacity to respond to the catastrophic in meaningful or effective ways. Thinking about the work of law in an age before the generalization of actuarial thinking and its technologies of containment, the chapter considers the long and always uneasy past of liberal hopes for law in what Samuel Beckett has marked as catastrophe's "long pause" afterward.[35]

"Here I Am!": Legal Existence Across the Abyss

Few commentators have devoted themselves as systematically to the problem of law and the self in the wake of violent political upheaval as novelist Honoré de Balzac. Although Balzac specialists tend to stress the author's conservatism, monarchical loyalties, and antagonism toward the leveling effects of revolutionary change, Balzac's fiction also gives voice to postrevolutionary liberal legal desire in the space of its illiberal critique.[36] Balzac's short novel, *Le Colonel Chabert*, first published in serial format in 1832, speaks especially powerfully of law's belated and uncertain confrontation with the aftermath of political catastrophe as it extended into the era of the Bourbon restoration.[37] A narrative of the seductive but impossible promise of law as remedy for catastrophe's injuries, *Chabert* explores both the instability of the boundaries that supposedly separate catastrophic moments from their longer term consequences and the repeated disintegration in the long pause afterward of the juridical categories and identities on which the meaningful operation of law claims to depend.[38]

The novel relates the story of a French army officer who falls on the field of battle in Prussia during Napoleon's 1807 campaign against Russia, is trampled by his own forces, declared dead, and then thrown, still living, into a mass grave. He manages to extricate himself from "the pit" (p. 1112) and, eventually, from a life of vagrancy, illness, and periodic institutionalization in Germany. Several years after the definitive restoration of the Bourbon monarchy in 1815, Colonel Chabert, Napoleonic count and "the hero of Eylau," returns to France unrecognizable and indigent.[39] Intent on reclaiming his legal identity, rewriting his civil status as living rather than dead, and restoring his rights to his wife and his wealth, Colonel Chabert single-mindedly seeks out the reparative powers of the law. As he puts it to the lawyer, Derville, "It seems perfectly simple to me. They thought I was dead. Here I am! Give me back my wife and my fortune   " (p. 1132)

Chabert's desire to return from legally declared death, his declaration of faith in the law to respond affirmatively to his announcement—"Here I am!"—so that he might be recognized and restored to himself, proves not to be so simple. Instead, Balzac's story insists on the ways public political catastrophe, collapsed into its lasting effects, resonates in the very possibility of coherent individual existence where it extends beyond biological survival. As it unfolds the story of the colonel's attempts to be present in the world of properly documented and juridically intelligible subjects, the novel meditates on the unsupportable legal paradox of the living dead in the liberal civic order of the early nineteenth century. At a deeper level, it explores the capacity of French civil law, itself the residue of revolutionary and Napoleonic upheaval, to secure the stability of identities in the aftermath of catastrophe.[40] As the critical Balzac will suggest, Chabert's assertion that he is "here," rather than inspiring juridical hospitality and the restoration of self, repeatedly threatens him with expulsion into the worlds of certified criminality or madness.

The story begins with Chabert's attempts to meet with his wife's attorney, Derville. In these opening pages, Balzac introduces Chabert as a juridical nonperson who returns to a France now officially devoted to closing "the abyss," a term of multiple meanings in the novel used in these early pages to signify the Revolutionary and Napoleonic eras, conflated into a single catastrophic rupture in the continuity of monarchical rule.[41] During his first visits to the lawyer's chambers in Paris, the colonel is repeatedly described in the terminology of the unfamiliar: He is "the stranger" (p. 1100), "*un chinois*," literally "Chinese," but also used to refer to something that promises to be excessively complicated (p. 1096), and as the clerks take in his cadaverous appearance, the "disinterred" (p. 1102).[42] When Chabert attempts to identify himself as a living man who is legally dead, he finds he is speaking an untenable contradiction, eliciting nonsense syllables and laughter from the law rather than its proper appreciation. "Pluff! Oh! Ah! Bàoun!" the clerks shout. "Trinn, la, la, trinn, trinn" (p. 1103). After he is at last admitted to see Derville and introduces himself as the Colonel Chabert "who was killed at Eylau," both the lawyer and his head clerk likewise assume that the paradoxical utterance cannot be true. The colonel's unsustainable assertion leads the men of law to conclude not that they are confronting the disintegrative consequences of catastrophic collapse but that they are faced with a delusional "madman" (p. 1110).

Chabert's dreadful unintelligibility, so comical to the unfeeling junior clerks and so clearly an indication of derangement for Derville, is of very particular sort. French law in the early nineteenth century specifically articulated a category of "civil death" for the living—that is, the metaphorical "death" of one's juridical person and the loss of the civil rights that attach to it. *La mort civile* was declared, however, only in the case of people facing capital punishment, forced labor in perpetuity, or deportation as a result of criminal proceedings.[43] In another section, the civil code anticipated the reappearance of a person formally declared missing and required the return of his property from those to whom the court had accorded its provisional or even permanent distribution.[44] The returned person or a legal representative, "armed with proof of his existence," would further be empowered by the law to challenge any subsequent union entered into by a spouse during the time of the absence.[45]

Neither of these regulated spaces of living death or protected disappearance facilitates Chabert's efforts to reclaim himself. Consigned by his legal demise to the ranks of the civil dead but never condemned, reappeared after a long absence but never formally declared missing, Chabert's presence in Paris is not ghostly but chimerical, a disturbing mix of physical and legal forms that elicits an uncontrolled flow of laughter or revulsion rather than belief.[46] In this way, Chabert occupies his own story as a juridically incoherent monster, disfigured by the confluence of personal and collective catastrophe, an embodiment of unexpectedly and unnaturally mixed categories who in turn brings forth a nonsensical cascade of mixed words and sounds from those who encounter him.[47] Thus, Balzac describes the clerk's reaction to Chabert's assertion of his impossible identity as "a torrent of cries, laughter, and exclamations, that would require all the language's onomatopoeia to depict" (p. 1103). A few pages later, Balzac returns to this theme, writing that Chabert's face was "something so dreadful that no human utterance could express it" (p. 1109).

The primary root of the colonel's problematic identity is documentary. The question of his existence depends entirely on the textual regime of *l'état civil*, or the state certification of a person's civil status, including the facts of birth, marriage, and death.[48] The authority of this documentary order to predicate meaningful existence should not be underestimated. Indeed, in the Napoleonic and post-Napoleonic liberal order, formal inscription in the municipal register constituted the very foundation of the civil subject who only then might exercise legal agency according to his or her gender, age, and family status.[49] His official death, the colonel explains to the lawyer, conformed to all the legal requirements: It was established by a "death certificate, likely prepared according to the rules of military jurisprudence" (p. 1111).[50] The documentary determination of Chabert's legal nonexistence recurs in his relations with his wife, now remarried to a rising noble of the old regime and reluctant to give up the advantages of her new life. The letters he wrote her during his exile in Germany, the duplicitous Madame Ferraud tells Chabert in their first meeting after his return, "were opened, soiled, the writing unrecognizable, and I had to believe, after having obtained Napoleon's signature on my new marriage certificate, that some clever imposter wanted to trick me" (p. 1155).

The project of reckoning with the documentary authority of *l'état civil* lies at the crux of Chabert's engagement with the law on the other side of the political abyss. It also structures his private conflict with his wife, who, Balzac notes, treads perilously close to a moral "abyss" (*abîme*) in her desperate efforts to avoid "precipitating" the personal and social "catastrophes" of marital breakdown and the revelation of her past as a courtesan (p. 1143).[51] Chabert thus describes his return from supposed physical death as an encounter with the weight of the documents that both maintained and depended on his metaphorical civil death:

> When I arose, dead, against a death certificate, a marriage certificate, and birth certificates, they [the lawyers he had consulted before seeking out Derville] threw me out, sometimes with that coldly polite manner that you put on to get rid of an unfortunate, sometimes brutally, as with those people who believe they are dealing with an imposter or a madman, depending on their characters. I was buried under the dead, but now I am buried under the living, under documents, under facts, under a whole society that wants me to return to the grave [*rentrer sous terre*] (pp. 1116–17).

Chabert despairs of the way his literal death can no longer be differentiated from his metaphorical death, a condition Balzac stresses throughout the story by repeatedly describing him as a living cadaver.[52] At the same time, the colonel's faith in the law to redress the disintegrating personal effects of political catastrophe persists for much of the novel, despite his documentary interment. He believes, at least during his early consultations with Derville, that his problems will be solved by the delivery of legal papers from Germany attesting to both his identity and his survival. These in turn would facilitate the correction of his *état civil* in France so that he might then bring suit as a legitimate legal agent against his uncooperative wife. As the story unfolds, however, Balzac suggests repeatedly that the documentary truth of the individual, carried by Napoleonic law across international borders as well as across the abyss from the Revolution to the Bourbon restoration, is far more difficult to stabilize than Chabert would believe. "It is a matter," Derville advises the colonel, "of proving [your identity] to people who have an interest in denying your existence. Thus your documents will be challenged" (p. 1132). "Everything," the lawyer reminds him at another point, "can be argued" (p. 1134). Even his wife's "innocent" bigamy, resulting in the birth of the children she never had with Chabert, will not be so simple to undo. "In your case," Derville contends, "the point of law is beyond the code . . ." (p. 1132).

The legal indeterminacy of the colonel's identity lies also in his genealogy. Abandoned at birth, Chabert's claim to his name, even when properly documented, rests on a fiction of parentage. "If I had any blood relatives," he laments, "maybe none of this would have happened, but, I must tell you, I am a foundling, a soldier whose patrimony was his courage, whose family was everyone, whose country was France, whose only protector was the good Lord. No, I'm wrong! I did have a father: the emperor!" (p. 1120). Chabert's horror of undocumented existence on the other side of the abyss that is the condition of his death brings back the condition of his infancy: Abandoned and *sans état*, he must rely in both cases exclusively on the state and its law to place him in the world of legitimate civil actors.[53]

Without the financial or emotion resources to "live life as a plaintiff" (p. 1135) while the documentary foundation of his identity is argued indefinitely, Chabert realizes as well that "compromise" in the form of an out-of-court settlement—the solution proposed to him by Derville—will not redress the injury of his painful hybrid condition. "Compromise? Am I dead or alive?" (p. 1123). Indeed, the text of the settlement drafted by the lawyer entails a contradictory blend of affirmation and renunciation. The countess Madame Ferraud, the settlement declares, will sign a notarized statement recognizing that Chabert is who he claims to be, ensuring that the registration of Chabert's legal death can be "voided." For his part, the colonel will renounce all marital rights and accept the formal annulment of his legal marriage.[54] In exchange for his agreement not to reclaim the property lost in the settlement of his estate at the time of his "death," finally, Madame Ferraud will pay Chabert an allowance of 24,000 francs. By these provisions, Derville's proposal aims to rectify Chabert's status as the living dead, but only by leaving the formal consequences of his civil death almost entirely in place, at least where they affect the organization of his past. Although the law will respond at last to his claim that he is "here" with a revision of his recorded death, the return to juridical life is nearly as empty as his legal death: Neither the return of his wife nor his fortune will follow from it.

The terms of the settlement nevertheless prove unacceptable to the countess, who sees the shadow of her own catastrophic social destruction behind any legal proceeding—that is, behind any public recognition of Chabert and, especially, behind the official revision of his recorded death. Rather than face Chabert in court, however, and rather than give up 24,000 francs of income, Madame Ferraud plots to ensure that Chabert's civil death becomes permanent by other means. It is at this point in the story that Balzac brings forward the alternative legal constructions of the colonel that have shadowed the identity he seeks as a rights-bearing, individuated civil agent returned from beyond the abyss: "Chabert" as motivated criminal imposter and "Chabert" as deluded madman.

Madame Ferraud works these alternatives simultaneously as she struggles to prevent the inauguration of her own catastrophe by reinforcing the colonel's unbearable civil death. First, through the elaborate staging of sentimental tableaus, she attempts to seduce Chabert into renouncing all he seeks in the settlement and admitting that he is an imposter out of pity for her and her children. At the same time, she colludes with her husband's secretary to have Chabert taken by force to Charenton, a notorious asylum that has haunted Chabert whenever his undocumented declaration that he was the dead war hero was met with laughter.[55] When he discovers each of these plans in turn, Chabert responds with horror, anger, and despair. Unwilling to sign "authentic" papers admitting in "crude terms" that he is an imposter and swindler, outraged by the plan to have him committed, and unable to bear the financial and emotional costs of an "odious" war of litigation with a wife who will not publicly acknowledge him, he finds that he no longer has the strength even to try to "jump the ditch" (p. 1163).[56]

The revelation of his wife's manipulation and betrayal strips Chabert of any hope that the law might organize his return. Exhausted, he momentarily considers suicide as the means of ending his bitter suspension between the legal worlds of the living and the dead (pp. 1162–64). Instead, the colonel decides to craft his own compromise. Secured by his word rather than by written text, the colonel's settlement rests on the promise to renounce the "illustrious" name of Chabert, although not by admitting he is an imposter or by submitting to the notion that he is deluded in claiming it. Rather, Chabert refuses proper naming as required by the registers of *l'état civil* altogether and declares that he will retain only his given name—that is, the foundling name that predates the Revolutionary and Napoleonic eras in which "Chabert" was made: "I am now only a poor devil named Hyacinthe . . ." (p. 1164).

Even that name loses its meaning in the final scenes of the novel after the colonel accepts his civil nonexistence and is "swallowed up" (*s'abîmer*) by "that mire of rags that swarms through the streets of Paris" (p. 1165). In these two episodes, the first set in criminal court where Chabert awaits sentencing for vagrancy and the second outside a state institution for the elderly decades later, Balzac stresses the ease with which even the self-declared identities of the undocumented collapse under the weight of the enumerative practices that organize the prosecution and incarceration of those without resources. "From the moment a man falls into the hands of justice," he comments on Chabert's vagrancy conviction, "he is nothing more than a responsible party [*être moral*], just as, for the statistician he becomes a number" (p. 1167). At Bicêtre, a hospital that also housed an old age hospice for the poor, he is identified by the narrator as "one of the two thousand" warehoused there in 1840 (p. 1169). Even Chabert embraces the ways identity beyond *l'état civil* gives way to the anonymous individuation of the number.[57] "Not Chabert! Not Chabert!" he shouts at Derville when the lawyer encounters him in the road by Bicêtre. "I call myself Hyacinthe . . . I am no longer a man. I am number 164, ward seven" (p. 1170).

Ironically, Chabert's quest for recognition within the framework of civil law ultimately leaves him abandoned within an institutional grid of individuated identification that denies him the status of human being. Unable to employ the law to ameliorate his monstrous juridical hybridity and help him "leap the ditch" of political and personal catastrophe, Chabert must in the end return to his prerevolutionary origins as a helpless dependent of the state. "What a life!" Derville comments. "Born in the foundling hospital, he returns to die in the old age hospice, after having helped Napoleon conquer Egypt and Europe in between" (p. 1171). Yet Balzac emphasizes Chabert's foundling condition, not to make him an exception but rather to make him exemplary.[58] Deprived of an extralegal identity embedded in relations of kin and orphaned in adulthood by the political exile of his "father," Napoleon, Chabert demonstrates both the fundamental documentary instability of state-regulated regimes of identity and the problematic way those regimes anchor legal agency in a civil sphere destroyed and rebuilt across the abyss.

Colonel Chabert concludes with Balzac's grim assessment of law's capacity to respond in any meaningful way to the desire for a secure civil identity and a just restitution of a legally meaningful self. While the priest might "purify, repair, and reconcile," Derville despairs in the last lines of the novel, the attorney's office is a "sewer that cannot be cleaned out." For Derville, the law can only bear witness to the worst in human beings, especially in their most intimate relations. Catastrophe appears in this conclusion in a doubled aspect: Public and political, the abyss in which "Chabert" came into existence, it is also mirrored for Balzac in the "moral abyss" of private and interior human existence after the Revolution. "I cannot describe to you what I've seen, for I have seen crimes against which justice is impotent," Derville tells Godeschal, a younger lawyer who is traveling with him when he comes across Chabert for the last time (p. 1172).

Balzac's concluding emphasis lies on the destructive excesses that surge from humankind's moral abyss and render law powerless to act or describe the unspeakable. As it insists on the rightful place of moral consideration in any reflection on political catastrophe, the novel also speaks directly to the double bar that obstructs the way to the full realization of a postrevolutionary, post-Napoleonic civil self as promised and—supposedly—secured by law. In writing the story of the colonel as a tale of foundational texts of identity that are simultaneously essential and also always open to dispute, Balzac challenges the dream of the abstract subject of civil rights as the survivor of the Revolutionary-Napoleonic era upon whom a new, well-ordered French society might be founded. In this regard, Balzac poses the ethical and juridical problem of the *sans-papiers*, or the undocumented person, for the rule of law in the modern West.[59] Although Chabert's entry into Restoration France is defined by the temporal-political rather than the geopolitical borders that tend to organize current debates about undocumented refugees and immigrants, he is nonetheless relegated to the status of those who claim France as physical and affective home but who cannot be placed in the textual order that secures meaningful legal existence and indigenous rights.[60]

Proper documentation is only one of obstacles blocking the way to the liberal fantasy of the rights-bearing civil individual who might anchor postrevolutionary society. Balzac also demands that the reader reckon with the way poverty creates unsurpassable barriers to the abstract individual agency presumed by civil law.

Without the means to pay for court fees or legal representation, the poor in *Colonel Chabert* are blocked from civil action and instead are ushered into the domain of law through the doors of presumed criminality. In this arena of transgression and its punishment, the anonymity of the undocumented person finds its modern bureaucratic response in the numbered registers of incarceration. The next section of this chapter takes up the theme of poverty, especially politicized poverty, as it bore upon the discussion of civil law, the legal self, and revolutionary catastrophe in the mid-nineteenth century.

1848: Catastrophe at the Threshold of Justice

Recalling his uneasy feelings about the "feverous and irregular symptoms" discernible in French political life shortly before the king was toppled by a revolutionary uprising in February 1848, Alexis de Tocqueville confessed that even he, one of the more "attentive observer[s]," was "far from supposing that catastrophe was so close and would prove so terrible . . ."[61] Although he had been no admirer of the monarchy as it had evolved in the 1830s and 1840s, Tocqueville could not contain his horror in the face of revolutionary chaos that appeared to him to violate his cherished ideal of "a regulated and orderly freedom, controlled by religious beliefs, mores and law."[62] Tocqueville turned repeatedly to the language of "catastrophe" to articulate his shock at the arrival of political events vaguely anticipated by some but "*unforeseen* by everyone."[63]

For Tocqueville, as other French liberals whose commitment to liberty was severely strained by their aversion to revolutionary violence and the democratic radicalism of the republican and socialist left, the February uprising was especially disturbing because of the ways it appeared to open the door to the antihierarchical and redistributive demands of politicized poverty. Here, too, the idiom of catastrophe, with its tropes of collapse, transgressive mixing, and the fundamental disturbance of linguistic order, penetrated Tocqueville's retrospective account of the dread he felt in the immediate aftermath of the February uprising:

> After the 25th February a thousand strange systems poured from the impetuous imaginations of innovators and spread through the troubled minds of the crowd. Everything except the Throne and Parliament was still standing; and yet it seemed that the shock of the revolution had reduced society itself to dust, and that there was an open competition for the plan of the new edifice to be put in its place; each man had his own scheme . . . One was going to abolish inequality of fortunes; another that of education; while a third

attacked the oldest inequality of all, that between men and women. There were remedies against poverty, and against that disease called work, which has afflicted man since the beginning of his existence.[64]

The destructive ideas that spilled and spread across Paris, threatening to lay waste to the familiar social order with their "extraordinary ferment and unheard-of disorder," all "adopted the common name of socialism."[65]

Tocqueville's account of revolutionary catastrophe as a leveling event that "reduced society itself to dust," thus producing a common ground in "socialism" for the alarming confluence of "a thousand strange systems," is valuable here not as a definitive history of the complex events of 1848 but for the way it crystallizes liberal anxiety about the ways popular uprising threatened to rewrite the future in unexpected ways. In particular, *Recollections* captures a widespread perception of calamitous destruction and the unstoppable collapse of once reliable boundaries that would come to inform liberal responses to the revolution and its more radical social claims, especially as those responses took shape in the longer wake of the new government's violent repression of popular insurrection in June.[66]

At the core of the liberal reaction to 1848 lay the concern about the revolutionaries' discursive relocation of the problem of inequality from the realm of individual civil standing before the law to the domain of social rights, including the right to work and the right to public assistance for those unable to work.[67] Those ideas, one influential opponent wrote in 1850, were monstrous, mixed from categories that should not be joined. They were "chimerical, ruinous, antisocial [and] sterile." The project of rebuilding in the aftermath, particularly where state aid was concerned, thus entailed the reimposition of proper boundaries, "the separation of the true and the false, the possible and the impossible."[68] Even after popular protest in defense of the right to work had been silenced by the army in June, and even after a new constitution redefined those newly asserted rights as matters of state voluntary "obligation" in late 1848, a move that supplanted the rights-bearing social citizen with the social and moral value of government "charity," liberal thinkers brought their fears of revolutionary catastrophe to a new round of reflections on law, poverty, inequality, and the stability of the civil self.[69]

This postrevolutionary liberal concern about political catastrophe, law, and the restoration of the civil self as the primary figure in any legitimate discussion of justice refocused attention on an issue that had been introduced in the French parliament in February 1848, just eleven days before antigovernment protestors took to the streets: whether France should legislate some form of state-provided legal aid for the poor.[70] At first glance, the documentary record suggests that liberal commentators and legislators merely picked up in mid-1849 where they had left off in February 1848.[71] Indeed, some of the central discursive work entailed the re-animation of narrative and figurative practices that had organized the increasingly lively discussion of law, *l'assistance judiciaire*, and the poor individual in the years before the revolution.[72] Beneath the surface of this apparent continuity, however, lay the enduring effects of a double catastrophic disturbance. For no one returning to this ground in 1849 could forget the "unanticipated" events that had so dramatically disrupted the rhythms of legislative process in February 1848. Nor could they suppress the fact that the theoretical and political ground for imagining legal equality had unexpectedly shifted beneath their feet. This seismic movement inspired both a historical revision and a fearful anticipation of the revolutionary future—simultaneously predictable and unknown—that would pervade *savant* reflection and parliamentary debate into the early 1850s. In this context, state-provided legal aid returned to the French liberal imagination as a critical project for repairing and preventing catastrophic social violence. At the same time, however, in its recognition of poverty as a meaningful category of analysis with regard to constitutional equality, legal aid threatened to transform the courtroom from a site where individual agents sought justice as supposed equals to a place where class-based passions might turn the civil suit into the building block of a new revolutionary barricade.[73]

Before February 1848, the question of public legal aid for the poor had been a matter of great interest for members of one of France's most influential bodies of liberal thinkers: the Academy of Moral and Political Sciences.[74] In 1847, Gustave de Beaumont, who had accompanied Tocqueville to America in the early 1830s, and Alexandre Vivien, a prominent member of the Academy's section on legislation whose work on legal aid would enter directly into parliamentary deliberation, presented their respective reports to their colleagues on how best to provide legal assistance to the French poor.[75] Although the reports engaged in spirited discussion of alternative structures of state aid—especially the comparative virtues of the *avocat des pauvres*, a state-appointed lawyer for the poor as found in parts of Italy, and judicial aid that provided fee relief rather than state-provided representation—their significance here lies in the academicians' larger vision of poverty, the civil individual, and the possibility of an equalized justice that made those alternatives thinkable solutions on the eve of the unforeseen "February catastrophe."[76]

At the heart of that larger vision was the characterization of the poor civil subject as impotent, vulnerable, and unable to overcome the barrier of fees and costs that obstructed his access to justice. Invoking France's reputation as an "advanced civilization," de Beaumont noted that justice was ever more "inaccessible" for the poor. "Strange progress," he remarked, "that incessantly perfects the protection of the strong, while making the support requested by the weak more and more impossible."[77] Vivien likewise read poverty before the law as a matter of individuated vulnerability. To render the alarming defenselessness of the impoverished civil person more tangible, and to develop support for his proposal in the stimulation of moral outrage, his report sketched melodramatic scenarios that tied sections of the civil code to the conventions of domestic fiction.[78] "If a wife is beaten by a brutal spouse . . . does not an interest in public safety require the declaration of legal separation? If a child born of adulterous relations imposes himself on the [legitimate] household, does not family order justify legal disavowal? If an ungrateful son refuses food to his needy father, does public morality in no way protest against this callowness?"[79]

To give even more poignant contours to his representation of the impoverished civil individual as unprotected from immoral violence—a vision that assimilated the universal male subject of law to the status of vulnerable dependent and the family threatened by "dangerous" illegitimacy to the position of the legal individual—Vivien drew upon the languages of misery and infirmity. The state, he contended, must recognize and respond to "the suffering of the poor individual, who, for lack of money, cannot exercise his rights . . ."[80] The poor person "who wants to claim his rights," he added, "is halted by obstacles that deny him the benefits of the common law and strike him with a kind of incapacity."[81] Vivien's emphasis on poverty as a source of both suffering and disabling constraint prepared the ground for his representation of state-provided legal aid as a philanthropic debt to constitutional equals whose active legal agency was compromised by want.[82] "Improve the condition of the lower classes," he concluded, "call upon them to participate in all the benefits of civilization, increase their well-being, suppress the inequalities that can be repaired [*corrigées*] by laws, that is the aim that we must set for ourselves . . ."[83]

In the liberal discourse of legal assistance before 1848, the poor civil subject thus emerged as a figure of weakness that could not enter fully into the law, or the civilization of which law was the emblem, without some form of charitable public intervention. He, for the civil individual who was incapacitated by poverty was generally represented here in terms of compromised masculine autonomy, was defined primarily by defenselessness, especially in the face of the lawless behavior of others, particularly in the domain of family and household. He was also rendered in terms of his inability to overcome the economic obstacles that barred access to a law celebrated precisely because it could not recognize the distinctions of wealth that in practice excluded him from its purview. In this way, the poor civil subject of pre-1848 liberal imagination occupied a paradoxical position: Equal by defini-tion, at least when the universal subject of law was taken to be the adult male, he was also always unequal, always in need of external supplement that simultaneously acknowledged and denied the difference poverty introduced into civil status.[84]

Despite the consistency with which they represented the poor civil subject as the fragile and suffering object of the state's charity, however, de Beaumont and Vivien could not suppress their concerns that the poverty might also enter the courtroom as antagonistic excess. In keeping with Tocqueville's retrojected antici-pations of revolutionary upheaval, their 1847 reports on legal assistance also bear traces of concern about "feverous and irregular symptoms" that might turn state aid into the tools of civil war. De Beaumont, for example, interrupted his study of the *avocat des pauvres* with a warning against the apparently increasing allure of radical solutions to economic inequality. "Rather than seek the remedy for social problems in very frightening and vague theories," he counseled, "many innova-tors, as impotent as they are well-intentioned, would do better to study what al-ready exists in different nations and look for those salutary things that might be borrowed from them."[85] Vivien issued an even stronger warning about legal aid's potential to facilitate catastrophic social conflict by opening the door to danger-ous admixtures of poverty and passion or by inverting social inequalities rather than ameliorating their effects. "The judicial arena," he declared, "must not be turned over to bad faith, to the spirit of vengeance, or to cupidity, which, under cover of indigence, would try to take the father, the landlord, or the honest and irreproachable citizen to court." Overempowering the poor with dangerous forms of legal assistance—for Vivien, the powers derived from free representation by a state functionary—would be the same as providing the "have-nots" with "hunting rights on the haves."[86]

In the long aftermath of 1848 and the violent June repression of popular protest, liberal jurists picked up the interrupted project of developing and pre-senting legislative proposals that would make legal aid for the poor a formal gov-ernment obligation. In some respects, the postrevolutionary discussion of legal assistance appeared to return to the narrative strategies and figurative choices that had organized earlier debate as if no "catastrophic" disturbance had intervened.

For Monsieur Bioche, a specialist in civil and commercial procedure and lawyer at the Paris Court of Appeals, the provision of legal aid still functioned as an index of national civilization. "In the laws of all civilized peoples, whether ancient or modern, there are special dispositions that facilitate poor people's access to the courts," he argued in the introduction to his 1849 proposal for legislation on *l'assistance judiciaire*.[87] In the same vein, Monsieur Delapalme, *conseiller* at the highest court of appeals, declared that the state provision of aid corresponded to "progress" in "custom and generous opinion . . ."[88]

Constructions of poverty as an obstacle blocking the way to justice and the lexicon of equal access also seemed to reenter liberal discourse uninflected by revolutionary conflict. Antoine de Vatimesnil, reporting to his colleagues in the National Assembly in 1850 on a draft bill dedicated to legal assistance, argued that "justice must be accessible to all . . ."[89] For Bioche, "the enormity of court costs" was best represented as an "obstacle" that was one of the "great difficulties for the poor person."[90] Delapalme developed the motif of the "barrier" throughout his discussion of state-provided aid: "there are so many barriers placed at the entrance of the temple of justice, barriers that the poor man cannot always clear."[91] For Delapalme and the other jurists, the poor individual was reinscribed as a figure of weakness whose suffering was best read through the lens of philanthropy. "Oppressed, powerless to shake off the yoke of that oppression, he discovered that he needed a key made of money to make the sanctuary of justice, where he was hoping that he would find some help, open for him."[92]

What should be made of such seeming discursive continuity across the revolutionary "abyss" in the consideration of legal assistance for the poor? One possibility is that the upheaval of 1848 did not substantially shake liberal confidence that providing equal access to the law through state-based philanthropy would ameliorate any suffering associated with economic inequality. To be sure, that confidence seems to have been robust enough to lead to the successful enactment of a law in 1851 that allowed the state to waive costs for those poor individuals whose means were deemed insufficient and whose causes were adjudged to be sound by a panel attached to the court.[93] Yet the silence on the turmoil of the revolutionary uprising, and especially on the central place of politicized poverty and the claims to assistance as a right that the citizenship of poverty had authorized in 1848, might also be read as a sign of extreme unease. For as the Bourbon restoration had attempted to efface the revolutionary and Napoleonic eras while simultaneously responding to the desires of liberal law that had been legitimated during that radical rupture in monarchical time by writing the civil code into its own charter, so too the liberal jurists of the mid-nineteenth century attempted to erase the revolutionary moment while amplifying the importance of law that would both speak to and mute poverty's radical pressure on the universal promise of equality. From this perspective, the liberal project of providing legal aid to the poor entailed retrospective revision and forward-looking reconstruction but could not acknowledge the destructive event as a legitimate moment in its history.[94]

The desire to close the gap over 1848 is especially evident in the way liberal jurists organized their stories of legislative precedent. De Vatimesnil, for example, told a tale of accreted steps toward the comprehensive provision of legal aid that began with Napoleonic decrees on pro bono representation and ended with an 1846 statute that waived the costs related to acquiring copies of civil status records for poor couples who wished to marry.[95] In 1850, he contended, the draft bill on *l'assistance judiciaire* did not entail any sort of radical change. Instead, "it was only a matter of regularizing by complete provisions that which already exists, whether in law or in fact."[96] Delapalme traced the antecedents to the law even further back, beginning his narrative of gradual progress in Roman law. A 1610 ruling from Henry IV's council of state, he noted, "contain[ed] the complete development of the principle of public assistance . . ."[97] For Bioche, the line between the 1849 legislative project and prerevolutionary calls for reform, especially the reports generated by the Academy of Moral and Political Sciences in 1847, was direct and uninterrupted.[98]

Yet the excesses that contemporary liberal commentators attributed to the revolutionary moment, and especially to the socialist radicalization of poverty, could not in the end be contained in juridical discourse by figurative consistency and genealogical continuity. The double catastrophic disturbance that brought deliberation on liberal models of legal aid to a halt in February 1848 and shook public assistance loose from the secure conceptual foundation of voluntary philanthropy often spilled through the discursive barriers erected by the jurists in the aftermath of the uprising. In these moments, it saturated into their pronouncements about progress and charity with the fear that the proposed measure, designed to clear the poor individual's pathway to the pursuit of justice as an equal, might in practice open the courtroom door to the violence of a suppressed but unfinished civil war.

In Delapalme's report, for example, the survey of precedent lingered on the way revolution had brutally interrupted the development of charitable projects for ameliorating the civil inequality of the poor, not in 1848 but in the late eighteenth century. In 1787, he explained, *l'Association de bienfaisance judiciaire* had been founded by "enlightened citizens" and "men of good will" who devoted themselves to rectifying "the unequal struggle of weakness against strength, or of poverty against wealth [*opulence*] . . ."[99] With the arrival of revolution in 1789, the work of the association was brought to a shockingly abrupt halt. Many of its members, Delapalme claimed, were killed by revolutionary violence, including the association's president, "the protector of the weak and oppressed," sentenced to death by the Jacobin revolutionary tribunal during the final throes of the Terror.[100] Delapalme's long detour into the effects of revolutionary violence on men of good will and their efforts to protect the poor legal subject makes audible his abiding concern about the destructive capacity of revolutionary catastrophe. By transposing recent anxieties about the return of a bloodthirsty popular dictatorship onto the historical past, Delapalme reproduced the recent effects of political catastrophe in his own text, interrupting the scholarly sequence of precedent with scenes of violence and implying by unspoken analogy a terrifying alternative fate for midcentury "men of good will" if the radicalism of 1848 had been permitted to escalate any further.[101]

In de Vatimesnil's report to the National Assembly, the project of removing the barriers that kept the poor individual from seeking justice as a civil equal also generated scenarios of unsettling conflict, the reversal of social hierarchy rather than the amelioration of its effects, and the unloosing of dangerous sentiment. If the legislator "makes it too difficult to obtain assistance," de Vatimesnil worried, "he runs the risk of stifling legitimate claims . . ." On the other hand, if he "opens the door too wide, he will harm both the interests of the treasury and those of the people against whom those who receive aid would bring suit." At both extremes, the provision of legal aid "would degenerate into injustice and the oppression of others; it would exhaust the sources of public revenue; and it would become disastrous [*funeste*] nourishment for harassment and litigiousness."[102] Should the proposed legislation authorize the creation of a state-salaried *avocat des pauvres*, he added, it would "accord the poor a formidable advantage over the rich . . ."[103] In this account of the calamities that legal aid might usher across the courtroom threshold, the poor and suffering individual who had stood at the center of liberal endorsements of legal assistance was thus transfigured by the enduring fear of social aggression across the divide of wealth. Excluded from justice by stringent requirements, excessively empowered, or inappropriately stimulated, the suffering individual incapacitated by poverty who inspired de Vatimesnil's report disappears into the collective menace of "the poor," a predatory collectivity still possessed, as Tocqueville wrote in his description of the "socialist" residue of the revolution, by "greedy envious desires."[104]

Class war was not the only danger imagined by the advocates of legal aid in the aftermath of 1848. De Vatimesnil also wrote revolutionary catastrophe into his anticipations of how the expansion of the state's judicial bureaucracy might create an additional scene of conflict, especially if the legislature adopted the model of the state-appointed *avocat des pauvres*. The effort to prepare children for careers in this expanding sphere of state service, he argued, would "give birth to the spirit of intrigue devoted to the attainment of the object of one's ambition, and, when that fails, the spirit of armed faction [*faction*] that would upend society and capture the desired position through disorder and violence."[105] Not only might the wrong form of legal assistance to the poor individual open the doors of the civil court to a dangerous poverty that had been executed, exiled, and incarcerated at the end of the June insurrection, but it also threatened to initiate ruinous antagonism among the educated and upwardly mobile, a civil war of civil servants.

The final version of the draft bill, approved by the National Assembly in January 1851, limited legal assistance to fee relief alone. Even without the dreaded magistracy of the poor, some liberal-minded observers believed that state-provided aid would not assure the restructuring of a France so recently rent by revolution into a world of appropriately restrained civil individuals. As one member of the legislature put it, unless defendants were also offered fee relief, legal aid to the poor would become "a means of disturbing one's neighbors' tranquility and attacking their wealth. In effect, this is the danger of the law. When a man receives assistance . . . he can attack his neighbor, force him to bear formidable expense, disturb his tranquility and diminish his fortune, without running any risk himself . . ."[106] Despite this final representation of legal aid's worrisome potential to reanimate revolutionary catastrophe case by case rather than street by street, the law was passed without further revision.

By 1851, then, the liberal juridical imagination had been deeply imprinted by the desire to preserve French civilization and its moral order by disaggregating the poor and returning them to a rule of law that simultaneously constituted, liberated, and contained the legal individual. At the same time, that imagination had also been suffused by the fear that state-based programs to encourage and regulate that return might "open the door too far": The unimpeded legal access of which liberal jurists dreamed was shaded by their worry about what or who would enter the courtroom under the state-supplemented sign of poverty. Where Balzac had captured the concern that civil individuals might lose their identities in the aftermath of political catastrophe and be swallowed by the mire of the streets, the jurists of the mid-nineteenth century feared that the indigent civil individual might all too easily bring the radicalized masses of the streets into the courtroom with him.

Not coincidentally, much of the parliamentary debate on the final draft of the law, and much of the juristic writing in the decades after it was enacted, focused on anxieties about both the indeterminacy of "poverty" in the operation of the law and the fraudulent declarations of indigence that indeterminacy threatened to invite.[107] Years after the suppression of its most radical incarnation in June 1848, poverty seemed ever more dangerous and opaque to liberal observers; as a category in the allocation of civil justice, it offered cover under which malicious sentiment might exploit the state's aid and enlist the law in the war on one's neighbors' wealth and tranquility. It threatened civil justice from within rather than from without. Anxiety about imposture also shaped the finer details of the law's stipulations on the allocation of aid. Only those who had provided copies of official tax records and a certificate issued by the mayor of the town in which the supplicant resided, supporting a sworn declaration of indigence, could claim poverty as their condition. Underscoring Balzac's earlier vision of legal personhood as a fragile textual status, especially for the poor, the 1851 law made it even more difficult to conceal the ways liberal legal agency was predicated upon positive regimes of documentation and the gatekeeping operation of state bureaucracy.

The disrupted discussion of state-provided legal aid in the mid-nineteenth century thus reveals the ways in which political catastrophe—in rerouting narrative trajectories and reworking the crossing metaphoric lines of obstruction and access, strength, and vulnerability—put enormous pressure on French liberalism's ambivalence about its "cherished principles." In the wake of 1848, the promise of freedom was darkened by the nightmare of open competition among socialism's "thousand strange systems," by ruthlessly competitive career ambitions ignited by the dangerous expansion of government and burning fiercely in the heart of bourgeois families, and by the moral dangers of unrestrained litigiousness. It became more and more difficult to imagine a stable balance point between too much liberty and too little, between too many barriers to justice and too few. Political catastrophe and its aftermath also brought into the open liberal desire for practices of regulation and purification that might protect French civilization from the dangerous excesses of popular politics and the unnatural confusion or admixture of categories that seemed to accompany it. Above all, the desire to create and use law to correct the imbalance that poverty introduced in civil equality was shadowed by the anxiety that even the best intentioned legislation might reverse inequalities rather than ameliorate their effects, that liberal law might generate catastrophic upheaval rather than contain it.

In *Le Colonel Chabert* and across the midcentury discussions of legal aid, events encoded as political catastrophe repeatedly rattle the supposedly natural integrity of the civil individual. For liberal observers, catastrophe's upending and disintegrating effects undermined quotidian forms of ambivalence about law and liberty such that their frictions became increasingly difficult to bear. Together, the two stories of law and political catastrophe in the French nineteenth century suggest that while narratives of "return" to the law generally foregrounded those people destroyed or excluded by revolutionary violence and its aftermath, liberal desire for law also sought some path of return and renewal for itself, both in the realm of abstract ideal and within the narrative frame of the historical particularity of the moment. Between 1815 and 1850, that reconstitution of desire entailed the reassertion of law's restorative powers, especially in regrounding the civil person, in recuperating the past, and in repairing and reframing France's claim to forward-looking civilization within new political circumstances. At the same time, the restoration of those hopes for law also brought with it the reconfiguration of anxiety about law's place in the genesis of human-made catastrophe, including catastrophes of moral dimension that breached the imagined lines between the public world of the legal individual and the private domain of husbands, wives, and children. Renewed assertions of faith in law's power to regulate freedom were mixed with the whispering of deep discomfort about law's possible implication in the troubling alternative futures—futures populated by dreadful chimerical admixtures of the true and the false, the possible and the impossible—that revolutionary catastrophe and its aftermath had unrelentingly placed before the liberal mind's eye.

Notes

1. "Réunion d'experts à Kobe sur les tremblements de terre et la santé," *Communiqué de presse* OMS/8, 24 janvier 1997, http://www.who.int/archives/inf-pr-1997/fr/cpf97-08.html.

2. Médecins sans frontières (MSF), *Soins urgents en situations de catastrophe* (Paris: Hermann, 1979), 198.

3. Ibid., 199.

4. Ibid., 204.

5. Ibid., 207.

6. François Ewald, "Insurance and Risk," in *The Foucault Effect: Studies in Governmentality*, Graham Burchell et al., eds. (Chicago: University of Chicago Press, 1991), 198.

7. Larry D. Gaunt and Numan A. Williams, *Commercial Liability Underwriting* (Malvern, PA: Insurance Institute of America, 1978), 608.

8. Ibid.

9. The traditional approach to insuring against catastrophe relies on the reinsurance market enhanced in some countries by state-organized pools. On the rise of risk as a category of thought and policymaking, see François Ewald, *L'État providence* (Paris: Grasset, 1986); and Robert Castel, "From Dangerousness to Risk," in *The Foucault Effect*, 281–98.

10. Jean-Marc Binon, "Le Contrat d'assurance dans l'Union européene," in *Encyclopédie de l'assurance*, François Ewald et al., eds. (Paris: Editions Economica, 1998), 990,

11. Armin Pröfrock, "Les Risques naturels catastrophiques," *Encyclopédie de l'assurance*, 1191–97; and Serge Magnon, "L'Indemnification des catastrophes naturelles en France," in *Encyclopédie de l'assurance*, 1208–11. The year 1997 saw the first large-scale issuing of "catastrophe bonds," a new kind of corporate financial instrument that supplements the reinsurance market and allows investors to take on the unlikely risk of catastrophe in return for an attractive rate of return on their investment principle.

12. Paul Veyne writes, "An event stands out against a background of uniformity; it is a difference, a thing we could not know a priori . . ." Veyne, *Writing History: Essays on Epistemology*, Mina Moore-Rinvolucri, trans. (Middletown, CT: Wesleyan University Press, 1984), 5. The catastrophic event, Veyne's statement suggests, might usefully be thought of as the most eventful of events.

13. *OED Online* (2nd ed.), http://www.oed.com/, s.v. catastrophe. Where the connection between catastrophe and drama is noted, French dictionaries cite the sixteenth-century use by Rabelais: "the end and the catastrophe of the play [*comédie*]." See, for example, Emile Littré, *Dictionnaire de la langue française* (Paris: Hachette, 1885–1889), s.v. catastrophe. The centrality of abrupt change also emerges in the mathematics of catastrophe theory developed by René Thom. As Carter Scholz explains in his short story, "A Catastrophe Machine," "In the austere vocabulary on mathematics, a catastrophe is not just a sudden turn of violence. It is a set of conditions under which steady change may cause abrupt effects. At some point in a war of forces, one gives way." Scholz, *The Amount to Carry* (New York: Picador, 2003), 32. On catastrophe theory more generally, including its applications in the analysis of human behavior, see Alexander Woodcock and Monte Davis, *Catastrophe Theory* (New York: E. P. Dutton, 1978); and P. T. Saunders, *An Introduction to Catastrophe Theory* (Cambridge: Cambridge University Press, 1980).

14. On the operation of the future anterior in narratives of origin, see Michael Fortun, "The Human Genome Project: Past, Present, and Future Anterior," in *Science, History and Social Activism: A Tribute to Everett Mendelsohn*, Garland E. Allen and Roy M. MacLeod, eds. (Dordrecht: Kluwer, 2002), 339–62.

15. In geology, *catastrophism* refers to a theory of change that stresses sudden interruption, or as Steven Jay Gould puts it, "intermittent paroxysms, often on a worldwide scale," rather than continuity or gradual evolution. Gould, "Toward a Vindication of Punctual Change," in *Catastrophes and Earth History: The New Uniformitarianism*, William A. Berggren and John A Van Couvering, eds. (Princeton, NJ: Princeton University Press, 1984), 13. See also Trevor Palmer, ed., *Controversy: Catastrophism and Evolution: The Ongoing Debate* (New York: Kluwer, 1999).

16. On the historical tension between scientific and religious explanations of natural disaster, see Alissa Johns, ed., *Dreadful Visitations: Confronting Natural Disaster in the Age of Enlightenment* (New York: Routledge, 1999).

17. Alexis de Tocqueville, *Recollections: The French Revolution of 1848*, George Lawrence, trans. (New Brunswick, NJ: Transaction Publishers, 1970), 61, 64.

18. Thanks to Michael Fortun for this insight. See also Drucilla Cornell, *The Imaginary Domain: Abortion, Pornography, and Sexual Harassment* (New York: Routledge, 1995), for the integrative authority of the imagined future, esp. 43–53.

19. This capacity for signifying excess may explain the term's place in languages of unself-conscious hyperbole as well as in camp or ironic practices of exaggeration, especially in the discourse of beauty or self-presentation. "What looks like a fashion statement on J.Lo or some homeboy rap singer," an article in an online edition of the Australian *Sunday Telegraph* announces, "is a fashion catastrophe on the streets of Sydney." Kate de Brito, "Fashion Heads Down the Wrong Trackies," *The Sunday Telegraph*, September 7, 2003, http://www.sundaytelegraph.news.com.au/. In his discussion of the history of cosmetic surgery, Sander Gilman quotes a prominent surgeon's description of a patient: "in short, that nose constituted a catastrophe in itself." Gilman, *Making the Body Beautiful: A Cultural History of Aesthetic Surgery* (Princeton, NJ: Princeton University Press, 1999), 168. On the hyperbolic use of catastrophe, see also the example provided by *Le Grand Robert*: "His last film was a catastrophe." *Le Grand Robert* (2nd ed.), s.v. catastrophe.

20. The problem of excess and the limits of representation may explain why the Holocaust is so often taken as the self-evident and paradigmatic catastrophe of human making. See, for example, Moishe Postone and Eric Santner, eds., *Catastrophe and Meaning: The Holocaust and the Twentieth Century* (Chicago: University of Chicago Press, 2003), esp. the editors' introduction.

21. In this vein, another authoritative French dictionary includes among the definitions of catastrophe, "event with particularly serious consequences even irreparable ones, resulting in ruin or disaster." *Trésor de la langue française: Dictionnaire de la langue du XIXe et XXe siècle (1789–1960)*, s.v. catastrophe.

22. Binon, "Le Contrat," in *Encyclopédie de l'assurance*, 990.

23. Scholz, *The Amount to Carry*, 32.

24. In the large literature on trauma and memory, see among others Bruno Bettelheim, *Surviving and Other Essays* (New York: Knopf, 1952); Dominick LaCapra, *Representing*

the Holocaust: History, Theory, Trauma (Ithaca, NY: Cornell University Press, 1994); Cathy Caruth, *Unclaimed Experience: Trauma, Narrative, and History* (Baltimore, MD: Johns Hopkins University Press, 1996); and Shoshana Felman et al., eds., *Testimony: Crises of Witnessing in Literature, Psychoanalysis, and History* (New York: Routledge, 1992).

25. On the constitution of the coherent subject through the law, see, for example, Drucilla Cornell's discussion of law and the "symbolic conditions of individuation" in *The Imaginary Domain*, 42–43.

26. In this respect, liberals confronting the unexpected admixtures created by revolutionary upheaval come to resemble Bruno Latour's moderns who both seek the separation and purification of categories and are complicit in the creation of the "impossible" hybrids and admixtures their insistence on purity denies. Latour, *We Have Never Been Modern*, Catherine Porter, trans. (Cambridge, MA: Harvard University Press, 1993). On liberalism's interest in the "technologies of security," see Thomas Osborne, "Security and Vitality: Drains, Liberalism and Power in the Nineteenth Century," in *Foucault and Political Reason: Liberalism, Neo-Liberalism and Rationalities of Government*, Andrew Barry et al., eds. (Chicago: University of Chicago Press, 1996), esp. 101–2.

27. The classic voices of late eighteenth- and early nineteenth-century reaction belong to Joseph de Maistre, Louis de Bonald, and Edmund Burke. For a discussion of these writers in relation to revolution, see Paul Beik, *The French Revolution Seen from the Right: Social Theories in Motion, 1789–1799*, Transactions of the American Philosophical Society, New Series, vol. 46, pt. 1 (Philadelphia: American Philosophical Society, 1956); and René Rémond, *The Right Wing in France from 1815 to de Gaulle* (2nd ed.), James M. Laux, trans. (Philadelphia: University of Pennsylvania Press, 1969).

28. Hannah Arendt, *On Revolution* (New York: Viking Press, 1962), 21.

29. Tocqueville, *Recollections*, 5.

30. Ewald, *L'État providence*, 47–140.

31. For a useful account of historians of the present and their attempt to disturb naturalized truths through genealogical inquiry, see Graham Burchell, "Peculiar Interests: Civil Society and Governing 'The System of Natural Liberty,'" in *The Foucault Effect*, 32.

32. Richard Terdiman describes Balzac's writing as "driven by a *negative* passion, to displace and annihilate a dominant depiction of the world" (emphasis in the original). Terdiman, *Discourse/Counterdiscourse: The Theory and Practice of Symbolic Resistance in Nineteenth-Century France* (Ithaca, NY: Cornell University Press, 1985), 12.

33. On the complexity of liberal identities and languages in nineteenth-century France, see Dudley Channing Barksdale, "Liberal Politics and Nascent Social Science in France: The Academy of Moral and Political Sciences, 1803–1852," unpublished doctoral dissertation, University of North Carolina at Chapel Hill (1986), xiv–xv.

34. M. N. S. Sellers argues that the "effect of the French revolution was to separate 'liberty' from 'republicanism' in the French imagination." Sellers, *The Sacred Fire of Liberty: Republicanism, Liberalism and the Law* (New York: New York University Press, 1998), 36.

See also André Jardin, *Histoire du libéralisme politique: de la crise de l'absolutisme à la constitution de 1875* (Paris: Hachette, 1985).

35. Samuel Beckett, *Three Plays: Ohio Impromptu, Catastrophe, What Where* (New York: Grove Press, 1984), 36.

36. On Balzac's engagement with France's revolutionary legacy, see René Alexandre Courteix, *Balzac et la Révolution française: Aspects idéologiques et politiques* (Paris: Presses Universitaires de France, 1997). For the details of Balzac's legal education, see Michel Lichtlé, "Balzac à l'école du droit," *L'Année balazacienne* (1982), 131–50.

37. Curiously, the act of reading Balzac has been figured more than once as itself a sign of dangerous movement or the precipitant of sudden plot reversals. See, for example, the social and moral collapse associated with reading Balzac (along with Chaucer and Rabelais) in Meridith Willson's *The Music Man* (1957). See also Dai Sijie's tale of forbidden books and their effects in China during the Cultural Revolution, *Balzac and the Little Chinese Seamstress*, Ina Rilke, trans. (New York: Knopf, 2001).

38. The tale first appeared in serial installments in the magazine *L'Artiste*, under the title *La Transaction*, between February 20 and March 13, 1832. For a full publication history, see the afternote by Henri Evans in *L'Oeuvre de Balzac*, vol. 1, Albert Béguin and Jean A. Ducourneau, eds. (Paris: Le club français du livre, 1966), xxv–xxviii. All citations unless otherwise noted come from the Béguin and Ducourneau edition. All translations, unless otherwise noted, are my own.

39. In this element of the story, *Le Colonel Chabert* resonates with older stories of soldiers returning unrecognized from war, including Homer's *Odyssey*, and, in France, the early modern tale of Martin Guerre. Peter Brooks argues that Balzac's story also stands as a variation in a long tradition of stories focused on people who are buried alive. Brooks, *Reading for the Plot: Design and Intention in Narrative* (New York: Knopf, 1984), 221.

40. Cathy Caruth has argued that *Colonel Chabert* is organized by the resurfacing of traumatic memory and the ways that the past, embodied both in the colonel and the civil code, haunts the present. Caruth, "The Claims of the Dead: History, Haunted Property, and the Law," *Critical Inquiry 28* (Winter 2002), 419–41. Caruth's systematic emphasis on trauma and history, and especially the heavy deployment of the metaphors of life, death, haunting, resurrection, and rebirth in her own analytical language, unfortunately also works to foreclose the possibility of developing a more historically rich discussion of code law and its place in post-Napoleonic French history. The code's adoption by the Bourbon monarchy, for example, rather than entailing the perpetuation of trauma, might more usefully be read as a calculated reuse of spolia from the ruins of a previous edifice that become integral to the new structure and its symbolic authority.

41. In describing the changing fortunes of a French nobleman whose lineage dated back to prerevolutionary days, Balzac writes that he was one of many in the early years of the Bourbon restoration "waiting for the abyss of the revolution to be closed . . ." (p. 1140). On "the abyss" as historical and political rupture, see also Caruth, "The Claims of the Dead," 433, n. 15.

42. Balzac underscores Chabert's enigma by repeatedly representing him as a stranger from the Orient, for example, as an "Egyptian" (p. 1131), a nickname for the soldiers who served in Egypt under Napoleon in 1799, as the "chinois," or as an unknown outsider entering France from the east when he returns from Germany. Since at least the seventeenth century, "egyptien" has also been used as a synonym for vagrant and provided the root of the derogative term *gypsy*.

43. Code Napoléon, Title I, Ch. II, art. 25, in *Les Cinq codes de l'empire français* (2nd ed.) (Paris: F. Guitel, 1812). All citations of the Code Napoléon come from this edition. Among the rights lost by virtue of *la mort civile* were the rights of property ownership, which passed to heirs "in the same manner as if he had died naturally and intestate"; the right to inherit property; the right to sue; and the right to marry. Existing marriages were also dissolved as if by natural death. The civil dead were further proscribed from acting as guardian or serving as witness, functions limited to adult men by other articles of the code.

44. Code Napoléon, Title IV, Ch. III: "Des effets de l'absence," esp. arts. 131–32. The law insisted that the missing person's absence be treated as provisional for a period of thirty years, or when one hundred years had elapsed from the date of the missing person's birth, a stipulation no doubt inspired by the decades of revolutionary and Napoleonic warring abroad. Even once the disappearance was considered definitive, the missing person who reappeared with proof of his identity would be entitled to the restoration of all that had been his. These articles presume an adult male subject fully endowed with property rights.

45. Code Napoléon, Title IV, Ch. II, art. 139.

46. Marcelle Marini discusses Chabert as a phantasm of heterogeneity and contradiction in "Chabert, mort ou vif," *Littérature* (February 1974), 92–112.

47. Jeffrey Jerome Cohen associates monstrous admixtures with the unsettling effects of "ontological liminality." Monsters, he argues "are disturbing hybrids whose externally incoherent bodies resist attempts to include them in any systematic structuration. And so the monster is dangerous, a form suspended between forms that threatens to smash distinctions." Cohen, "Monster Culture (Seven Theses)," in *Monster Theory: Reading Culture*, Jeffrey Jerome Cohen, ed. (Minneapolis: University of Minnesota Press, 1996), 6.

48. The secular regime of *l'état civil* was organized at the same time that the revolutionaries were abolishing the monarchy and declaring France a republic. Responsibility for recording births, deaths, and marriages was moved from the hands of the parish priest to the offices of municipal authorities. This shift was preserved in the Code Napoléon, Title II, arts. 34–101. Historian Isser Woloch argues that the history of soldiering, draft evasion, and imposture was closely entwined with the founding of the secular regime of *l'état civil.* "Hometown mayors often produced false birth certificates, attestations that conscripts had satisfied the conscription laws or internal passports—documents which allowed a *réfractaire* to begin creating a new identity elsewhere." Woloch, *The New Regime: Transformations of the French Civic Order, 1789–1820s* (New York: Norton, 1994), 132, 414. *Le Colonel Chabert* explores an ironic reversal of this pattern.

49. I differentiate here between a liberal civil order organized around a rights-bearing individual whose agency is located in the domain of private property, contractual relations, and family and the freedoms commonly associated with liberal democracy. Neither Napoleon nor the restored Bourbon monarchs were favorably inclined toward democratic politics or political liberties, least of all freedom of expression. On Louis XVIII's retention of the Napoleonic civil code in the Charter of 1814 and the importance of the rule of law in civil matters under the Bourbon restoration, see John Hall Stewart, *The Restoration Era in France, 1814–1830* (Princeton, NJ: Van Nostrand, 1968), 19–20, 31–32. See also Guillaume de Bertier de Sauvigny, *La Restauration* (Paris: Flammarion, 1955), 98. On the code's strict limits on women's status as legal individuals, especially married women, see Claire Goldberg Moses, *French Feminism in the Nineteenth Century* (Albany: State University of New York Press, 1984), 18–20. See also Marie Henriette Faillie, *La Femme et le code civil dans la Comédie humaine d'Honoré de Balzac* (Paris: M. Didier, 1968).

50. On the military preparation of death certificates and their communication to civil authorities, see Code Napoléon, Title II, Ch. V, arts. 96–98.

51. On catastrophe as part of the historical lexicon of marital crisis, see Thomas E. Buckley, *The Great Catastrophe of My Life: Divorce in the Old Dominion* (Chapel Hill: University of North Carolina Press, 2002).

52. Antoine de Baecque notes the importance of the cadaver and its poetics in postrevolutionary French political culture when he argues that the effort to exhume the body of Louis XVI, executed in 1792, was one of the founding projects of the restored monarchy in 1814–1815. de Baecque, *Glory and Terror: Seven Deaths Under the French Revolution*, Charlotte Mandell, trans. (New York: Routledge, 2003), 116–17. On collapsed metaphor in *Le Colonel Chabert*, see Brooks, *Reading for the Plot*, 223–29.

53. On the treatment of abandoned infants in the later eighteenth century, see Albert Dupoux, *Sur les pas de Monsieur Vincent: Trois cents ans d'histoire parisienne de l'enfance abandonnée* (Paris: Revue de l'Assistance publique à Paris, 1958), 47–126. The well-known high mortality rates at the foundling hospitals of this period—some over 90%—suggest that Balzac locates his protagonist's very origins in the escape from near-certain death. "I emerged from the belly of the pit," Chabert tells Derville, "as naked as when I emerged from my mother's belly" (p. 1114).

54. Balzac's use of the term *void* [*annuler*] is curious here. Although the civil code allows for the correction of the registers of *l'état civil*, it requires that all modifications be entered separately and that "a note must be made in the margin of the emended [*réformé*] document." Code Napoléon, Title II, Ch. V, art. 101. The settlement thus could leave a secondary layer of textual revision but could not eliminate the recorded death.

55. In Germany, Chabert tells Derville, "they laughed in the face of the man who claimed to be Colonel Chabert. After a while this laughter, these doubts, got me worked up into a terrible state and got me locked up as a madman in Stuttgart" (p. 1115). Chabert realizes that the declaration of his paradoxical identity might also lead to the asylum in France: "When I told him that I was Colonel Chabert, he began to laugh so openly that I left without a word.

My detention in Stuttgart made me think of Charenton" (p. 1122). On Charenton during and after the revolution, see Jan Goldstein, *Console and Classify: The French Psychiatric Profession in the Nineteenth Century* (Cambridge: Cambridge University Press, 1987), 113–16.

56. Balzac organizes Chabert's discoveries of his wife's perfidy around the topography of her country estate. Although the ditch is there to be jumped literally, as Chabert unexpectedly reveals himself to his plotting wife, it also functions metaphorically as the double of the abyss.

57. On the development of enumerated individuation in the institutional organization of space, time, and identity, see Michel Foucault, *Discipline and Punish: The Birth of the Prison*, Alan Sheridan, trans. (New York: Vintage, 1979).

58. As both orphan and childless adult, Chabert may also give form to Balzac's bitter vision of liberal legal desire in the nineteenth century: abandoned by its revolutionary progenitor and without heirs that might fulfill its promise in the future. Thanks to Michael Dintenfass for suggesting this reading.

59. Gerard Noiriel treats the legal and bureaucratic regimes of documentation and the *sans-papiers* in *Réfugiés et sans papiers: La République face au droit de l'asile, xixe-xxe siècle*, first published under the title *La Tyrannie du national: Le droit de l'asile en Europe (1793–1993)* (Paris: Calman Levy, 1991), esp. 156–245. See also D. Fassin et al., eds., *Les lois de l'inhospitalité: les politiques de l'immigration à l'épreuve des sans-papiers* (Paris: La Decouverte, 1997). On the history of the passport as an internationally recognized identity document, see John Torpey, *The Invention of the Passport: Surveillance, Citizenship and the State* (Cambridge: Cambridge University Press, 2000).

60. The problem of chimerical mixing is evident in the legal and bureaucratic discourse of the *sans-papiers* in the late twentieth century. Thus, Mireille Rosello observes that since the immigration law of 1993, "the status of many immigrants has changed, turning some situations into inextricable nightmares, creating administrative monsters such as the *inexpulsable-irrégularisable* who can neither be deported nor given resident's status." Rosello, "Representing Illegal Immigrants in France: From *clandestins* to *l'affaire des sans-papiers de Saint-Bernard*," *The Journal of European Studies* 28 (1998), 138.

61. Tocqueville, *Recollections*, 11.

62. Ibid., 65.

63. Ibid., 64. See also pp. 7, 17, 23 for Tocqueville's reliance on this term. In a speech delivered to the Chamber of Deputies on January 29, 1848, Tocqueville also employed the image of the abyss to figure his sense of impending political calamity: "[F]or God's sake," he exclaimed, "change the spirit of the government, for, I repeat, it is the spirit that is leading you to the abyss." *Recollections*, 15.

64. Tocqueville, *Recollections*, 74.

65. Ibid., 74, 75.

66. On conservative and liberal fears about the June insurrection as a "communistic" uprising, see Jonathan Sperber, *The European Revolutions, 1848–1851* (Cambridge: Cambridge University Press, 1994), 199–200.

67. See article 2 of the draft constitution presented to the National Assembly just days before the June massacres: "The constitution guarantees to all citizens: [the rights of] liberty, equality, security, education, work, property, and assistance." Paul Bastide, *Doctrines et institutions politiques de la seconde république*, vol. 2 (Paris: Hachette, 1945), 291. The right to work had been established within days of the February uprising by a decree issued by the provisional government.

68. Adolphe Thiers, *De l'assistance et de la prévoyance publique* (Brussels: C. Muquardt, 1850), 7–8. Later in the text, Thiers describes both the socialist critique of capitalism and its counterideal of work as a right as "perfectly chimerical." Thiers, *De l'assistance*, 47.

69. By the end of the June Days, an estimated 3,000 protesters had been summarily executed by government forces of order. Another 12,000 were placed under arrest. More than a third of those arrested were later deported to Algeria. Sperber, *The European Revolutions*, 199. For the left, it was the government's June massacre that constituted the "catastrophe" of 1848. On the language of obligation in the constitution of the Second French Republic approved by the largely liberal National Assembly on November 4, 1848, see Bastide, *Doctrines*, vol. 2, 325–26.

70. France, *Le Moniteur universel* (hereafter *MU*), "Mémoire sur la défense des indigents dans les procès civils et criminels," February 11, 1848, 357–358. Vivien's report, followed by commentaries by several other members of the Academy, appeared the previous year in the Academy's publication, *Compte rendu des séances et travaux de l'Academie des sciences morales et politiques*, 2 (1847).

71. Maurice Agulhon contends that June 1849 marks an important moment in which the conservative reaction to 1848 was consolidated, a shift he locates particularly in the severe restrictions placed on political speech. Agulhon, *The Republican Experiment, 1848–1852*, Janet Lloyd, trans. (Paris and Cambridge: Maison des Sciences de l'Homme and Cambridge University Press, 1983), 119.

72. The 1840s also saw the rise of socialist claims on the scientific analysis of society, a political and conceptual challenge from the margins that pushed members of the Academy to stress the unique legitimacy of liberal reform. See Barksdale, "Liberal Politics," xi.

73. Giovanna Procacci describes the paradoxical legacy of the revolution of 1789 for liberals in similar terms: "It was in their capacity as the poor that they had to be integrated into the domain of the law, yet their manifest inequality could not contradict the equality that was established by law . . ." Procacci, *Gouverner la misère: La question sociale en France 1789–1848* (Paris: Seuil, 1993), 77.

74. On the history of the Academy and its work at the nexus of scholarly life and politics, see Barksdale, "Liberal Politics."

75. Gustave de Beaumont, "Rapport sur l'administration de la justice civile et commerciale en Sardaigne, suivi d'observations par MM. Charles Lucas, G. de Beaumont, et Cousin," *Compte rendu 1* (1847). Among his many projects, de Beaumont had collaborated with Tocqueville on a study of the American penitentiary system in the early 1830s and

wrote a novel about slavery in the United States, first published in France in 1835. Vivien, "Mémoire sur la défense des indigents," *Compte rendu 2* (1847).

76. Tocqueville, *Recollections*, 7.

77. De Beaumont, "Rapport," 25.

78. On nineteenth-century family fiction as a discourse of danger and order, see Roddey Reid, *Families in Jeopardy: Regulating the Social Body in France, 1750–1910* (Stanford, CA: Stanford University Press, 1993).

79. Vivien, "Mémoire sur la défense des indigents," *Compte rendu 2*, 462. For relevant family law on marital separation, paternity claims, and filial obligation, see Code Napoléon, Title VI, Ch. V, arts. 306–11; Title VII, Ch. III, art. 340; and Title V, Ch. V, art. 205.

80. Vivien, "Mémoire," *MU*, 358.

81. Ibid.

82. On the critical move in liberal discussions of poverty from politics to morality and philanthropy, see Procacci, *Gouverner*, 16.

83. Vivien, "Mémoire sur la défense des indigents," *Compte rendu 2*, 464.

84. See also Procacci, *Gouverner*, 85.

85. De Beaumont, "Rapport," 31.

86. Vivien, "Mémoire," *MU*, 358.

87. M. Bioche, "Proposition relative a l'institution de Bureau de l'avocat des pauvres," *Journal de procédure civile et commerciale 15*, 2ème série (1849), 89.

88. Delapalme, De l'assistance judiciaire," *Annales de la charité 6* (1850), 676.

89. "Rapport fait par M. de Vatimesnil, au nom de la commission chargée d'examiner le project de loi sur l'assistance judiciaire, et la proposition de M. Favreau," *MU*, November 26, 1850, 3364.

90. Bioche, "Proposition," 91.

91. Delapalme, "De l'assistance judiciaire," 667.

92. Ibid., 668.

93. For the full text of the law, see *Recueil Sirey: Lois annotées, 1848–54* (Paris: Administration du recueil général des lois et des arrêts, 1854), 10–15.

94. In his study of art and the Paris Commune of 1871, art historian Alfred Boime notes a similar tendency to revise the revolutionary past in the interests of a different future among Impressionists painting in the aftermath of that urban uprising and its violent suppression. Boime, *Art and the French Commune: Imagining Paris After War and Revolution* (Princeton, NJ: Princeton University Press, 1995), esp. 96.

95. De Vatimesnil, "Rapport," 3364. On the identification of early nineteenth-century legislation as the immediate antecedent of the draft bill under consideration in 1850, see also M. Blandin, "L'avocat des pauvres à Pau," *Annales de la Charité 6* (1850).

96. De Vatimesnil, "Rapport," 3365.

97. Delapalme, "De l'assistance judiciaire," 668, 671.

98. Bioche, "Proposition," 89.

99. Delapalme, "De l'assistance judiciaire," 673–74. Delapalme quotes the president of the association, Boucher d'Argis, when he refers to the "unequal struggle."

100. Ibid., 675–76.

101. François Furet explains the violence of actions against workers in June 1848 as the bourgeoisie's effort "to exorcise its retrospective panic of 1793 . . ." Furet, "The Tyranny of Revolutionary Memory," in *Fictions of the French Revolution,* Bernadette Fort, ed. (Evanston, IL: Northwestern University Press, 1991), 158–59.

102. De Vatimesnil, "Rapport," 3365. See also Bioche, "Proposition," 90: "To open the way to the courts for all of the poor without any guarantee of the merits of their claims, would be to multiply to infinity the causes of dispute and strife among citizens . . ."

103. De Vatimesnil, "Rapport," 3365. Delapalme underscored these positions in his essay, quoting liberally from de Vatimesnil's report where it touched on safeguarding both the state and "others" (*les tiers*) from any "unfair advantage" that state representation might give to the poor. "Assistance too liberally given," he wrote, again quoting de Vatimesnil, "would be a disastrous encouragement to the mania for litigation . . ." Delapalme, "De l'assistance judiciaire," 677, 678.

104. Tocqueville, *Recollections,* 165.

105. De Vatimesnil, "Rapport," 3365.

106. M. Defontaine in *MU,* 3rd deliberation, January 22, 1851, 232.

107. "2ème délibération des projets sur l'assistance judiciaire," *MU,* December 7, 1850, 3498–501. For later commentary, see among others Alfred Levesque, "De la loi du 22 janvier 1851 sur l'assistance justiciaire et des modifications qu'elle réclame," *Revue pratique de droit français,* vol. 2 (August 15, 1856–February 1, 1857).

Committed to Memory:
Rebecca West's Nuremberg

RAVIT PE'ER-LAMO REICHMAN

For the law, like art, is always vainly racing to catch up with experience.
—Rebecca West, *The Meaning of Treason*[1]

The banality of evil is arguable, the banality of boredom is manifest.
—Patricia Meyer Spacks, *Boredom*[2]

To imagine law's encounter with catastrophe is to tell a story of expectations—particularly, the expectation that law will make sense of catastrophe, that it will generate an understanding of its vicissitudes and a judgment of its horrors. But this *story* of expectations suggests, too, the narrative demands we bring to law: the story of our desire for a gripping, engaging encounter, one that makes sense because it takes place before an enchanted audience.

Yet the experience of law—and of law's response to catastrophe—is often quite another matter. Frequently, law's effort to cope with historical disaster yields anything but an exhilarating narrative. Often, it is simply dull. What does one do when the story of law and cataclysm falls short of expectations? What, in other words, do we do when the law leaves us bored stiff?

The unprecedented experience of the Nuremberg Trial posed these questions in a historical and pressing way, and it is from the perspective of this trial that I propose to answer them. For if World War I was experienced as an unprecedented military trauma, World War II extended this trauma into the juridical realm. With the establishment of the Nuremberg Laws in 1935, Nazi Germany created a legal basis for what has been referred to as an "administrative massacre"[3]—a legal system that would pave the way for the horrors of the Final Solution. After the war, the Nuremberg War Crimes Tribunal[4] sought to reassert a world where law could once more align itself with justice.[5] Held in Nuremberg's restored Palace of Justice, conducted in four languages, and presided over by eight judges, the trial was conducted in 403 open sessions from November 1945 until October 1946.

On trial were twenty-two defendants, all high-ranking Nazi officials representing a range of organizations, among them the SS, the Gestapo, the Reichsbank, and the German Armed Forces High Command.

While the lawyers and judges labored to make legal sense of the past and to leave a record of that past for the future, the 250 journalists assembled in the Palace of Justice worked to convey this unprecedented legal event to the outside world. Trials, Laurel Leff reminds us, hold a particular appeal for journalists, in that they contain "a satisfying narrative arc: they open, they close, with moments of drama (and tedium) in between. Most important, they give resolution in the form of a verdict. The Nuremberg trial had the additional allure of offering an explanation for the cataclysm the world had just endured."[6]

However, not all journalists relished the easily digestible format of the Nuremberg Trial's narrative arc. As a writer with no legal training, Rebecca West's reports from Nuremberg for *The Daily Telegraph* proved some of the most memorable and surprising, and it is in both senses that I propose to explore the essays as conveying an experience, rather than an explication, of law.[7] These essays represent, I believe, instances of legal writing that transform the trial from the unlikeliest of vantage points, offering a sense of what it felt like to bear witness to a moment of historical justice. Despite their innovativeness, they have been almost entirely overlooked by critics, who instead focus on West's epic, genre-defying work *Black Lamb and Grey Falcon* (1941).[8] I propose to restore to these pieces their remarkable potency and in doing so to explore the rhetorical and associative ways in which they shed light on Nuremberg.

Born in 1892, West lived through both world wars and wrote about each of them. In her prolific career, she wrote about historical events in a wide range of genres. The Great War serves as the subject of her first novel, *The Return of the Soldier*, which took up the issue of shell shock. By the end of World War II, however, West would no longer engage the past as an author of fiction, but as a witness to a still-emerging history. The change had as much to do with West's political interests as it did with the assignments she received from newspaper editors. And these assignments seemed to bring her, time and again, to courts of law. Following her stint at Nuremberg, West's journalistic role was to become increasingly legal. Despite having no formal training in law, she became the most renowned legal correspondent of her day, covering trials at home and abroad. And yet she was ambivalent, even somewhat resentful, of this role. In an untitled and incomplete autobiographical sketch, written presumably toward the end of her life—she died in 1983 at age ninety—West looked back at her work as a legal reporter with a generous measure of irony:

> After the war I went out and looked at the world for the New Yorker. I did this because I realised that the world had wholly changed—and I still think it is under-reported. The first thing I did was the trial of William Joyce, and then I went to the Nuremberg trials. Unfortunately I did a number of trials connected with treachery [. . .] because they threw light on why people become totalitarians, and the result was that ever since every body has tried to make me go and report trials. When a serious crime has committed [sic] people telephone for the police and the doctor and then the newspaper syndicates get on to me. This is unfortunate because there is nothing I loathe like trials and law-courts. I sit and weep in court, which looks mad, and get a headache—and I have such contempt for people who go to trials for fun that I hate to be present when they are. The misery behind any trial is so great that one would have to be a monster not to be appalled.
>
> Now I have settled down to writing great long novels and I intend to do nothing else until I die, whenever that may be.[9]

West's description of her own apparent madness during a trial, "I sit and weep in court," seems startling from a writer who, by sheer virtue of her frequent exposure to courtrooms, should have been accustomed to her share of courtroom drama. Yet her disavowal of law and her determination to return to literature point to her complicated emotional relationship to public justice, one that differs remarkably from the sensational swell of collective emotion that ordinarily accompanies a high-profile trial. What can we make of West's departure from the courtroom? What does her loathing of the law tell us about her experience of it? How can West's account of, and ambivalence about, law shed light on the expectations one brings to public manifestations of justice?

I would like to suggest an answer to these questions by examining West's sense of history and memory in her reports on Nuremberg. West's experience at the trial marked the beginning of her career as a legal journalist, charting a course away from her fictional writing about World War I and toward her non-fictional response to World War II. In *The Return of the Soldier*, her first novel, West confronted the Great War through a fictional world; by the end of World War II, however, she would no longer imagine this world in fiction but would witness it as a spectator and a correspondent for the *Daily Telegraph* at the Nuremberg Trial. I would like to think about the divergence of these approaches to war and history not merely as generic differences but in terms of West's changing sense of responsibility to history—a transformation that we find in her interwar prose and one that dramatically alters her approach to remembering and recording the past.

In her 1918 novel *The Return of the Soldier*, West infuses a traditional romantic plot—a man torn between two women—with the experience of shell shock. The novel tells the story of Chris Baldry, a soldier who returns from the front to a world he no longer recognizes. He remembers nothing of his recent past: His wife, Kitty, is a stranger to him, and the life he led at the lush Baldry Court before the war now seems sterile and foreign. Instead of this past, Chris's memory retreats fifteen years to the brief but powerful love affair he had at age twenty with a woman named Margaret, who has since married and moved with her husband to a nearby working-class town. When his wife Kitty and his cousin Jenny, the story's narrator, meet Margaret, they see a drab, lower-class woman whose homeliness makes it hard to imagine that she ever ignited passion in a man as sophisticated as Chris. The question of whether Chris will eventually recognize his present life (he ultimately does) provides West with a suspenseful and modern narrative twist on the traditional love triangle.

But if the Great War set the tone for the dramatic trajectory of West's first novel, her writing after World War II was decidedly less riveting. Instead of finding drama in Nuremberg—or creating it herself in her reports on the trial—she insisted on telling the story of its incredible dullness. The trial felt like "water torture, boredom falling drop by drop on the same spot of the soul,"[10] and no amount of literary ambition could prevail upon West to recount the trial differently by according it gripping proportions to spark her readers' interest.

Given the high stakes of the trial, it may have been wiser to sidestep this boredom altogether and to focus instead, as most newspaper articles did, on Nuremberg's sporadic moments of courtroom drama. Instead, West chose to convey the atmosphere of the court in all of its unbearable dullness. Why did she deliberately flatten an event of such magnitude? What had become of the Rebecca West who, in 1918, let World War I transform her literary lens, refracting the novelistic convention of a man choosing between two loves through the painful condition of shell shock? How, in short, do we account for the differences in the two works—differences that compelled West to write of World War I in a dramatic, fictional narrative but to record the trial of the Nazi criminals, presumably no less dramatic, with unparalleled matter of factness?

Demands of a Contemporary Age

The answer to these questions lies in West's changing sense of the past's relationship to the present—a sense that might best be described through her notions of contemporariness and relevance. She advanced these ideas most vigorously in her interwar essay "The Dead Hand," which explores the vicissitudes of memory instantiated in London's architecture. In this piece, West insisted that works of art—buildings, monuments, novels—should be integrated into everyday life rather than standing apart from it. Her point is not merely an aesthetic one but is rather a historical argument about the ills of turning away from the present. It is along these lines that West criticized London's architecture as being old, safe, and unrelated to the life around it. Walking through Piccadilly, she is struck by the city's indifference to the present:

> Throughout that walk I believe I thought of nobody who belonged to an age later than Albert the Good. Why should I? I pass no buildings which are not either old or designed to harmonize with old neighbours. Not one single architect has found it possible to erect anywhere in those two miles a façade that has any reference to modern life [. . .] In fact, when I am not feeling in a mood to stand up manfully to life, London enables me to pass in the drooping of an eyelid to a comfortable revisiting of the past; and I can disregard, and dislike for their invasions on my reverie, the people in the streets who happen to compose the age in which I live.[11]

These edifices stand out—and consequently stand off—as buildings that distinguish themselves as those of another era; they no longer bear any relevance to their times. And the same disavowal of modernity also extends to literature. West thus moves from the "comfortable revisiting of the past" in architecture to an analogous impulse in fiction, lambasting the novels being published (and much to her dismay, praised) in England at the time. "Now, why did these books enjoy the passionate favour of the English critics?" she asks. "The answer," she continues, "is simply that they contain nothing that is relevant to our age. They are blankly not contemporary. They might quite well have been written in 1899."[12]

Her explanation is odd indeed: Discerning critics, after all, would be expected to prize contemporary, relevant works over old-fashioned novels. West clarifies her counterintuitive stance by positing this reactionary critical practice as a consequence of the war and as a symptom of the postwar era. In the current literary climate, she proposes that the excessive praise heaped on second-rate novels issues not from personal whim or solipsistic taste but from the generational divide after the Great War.

There is now, due to the very slowly emergent consequences of the war, a very clean-cut division between young and old minds. The books which are liked by people under forty are, as a general rule, not the same as the books which are liked by people over forty; and this means that some of the older writers find themselves diminishing in importance far more rapidly than men who have arrived at such eminence have done during their lifetimes in any other age. They frequently try to arrest the landslide by tampering with our critical standards. They overpraise work done in the old manner (which is naturally followed by second-rate and timid minds) and underpraise work done in the new manner (which is naturally followed by first-rate and audacious minds).[13]

The questions of relevance and contemporariness, West implies, would not have become pressing if not for the war, which precipitated the generational rift that would see the old guard of critics attempting to stem the tide of new writing and thus unduly praising old-fashioned novels. This impulse, however, is not merely a product of nepotism or protectionism. Instead, West argues, it issues from the tendency to avoid a painful present by turning back to a more reassuring past. It is not, in other words, just a tug of war between generations that produces overpraise of bad novels. It is rather, as West sees it,

this English habit of wandering into the past as a refuge from the distressful present. There is a reason why this should be an English and not a generally European habit. The past we can escape to through our associations is not merely the past, it is peace. Between the Crimean and the South African wars nothing military vexed us save distant consequences of our militarist expansion; and at home we had a succession of steady governments. The same period in France was split across by the War of 1870: and from then until the Great War it had an average of a government a year. Why should the Frenchman exchange unrest for unrest by going back a couple of decades? [. . .] [T]he average French critic stays where he is and takes what comes; while the average English critic stays where he was and takes what used to come when he was a boy.[14]

Yet however ardently one wishes for the restoration of prewar life, this return only reinforces, with renewed strength, the division between past and present, old and young. And while it may be more comfortable to wander back into this past, this return comes at great social cost. For its irrelevance to a modern, difficult present does more than shape bad taste in novels: It prevents genuine engagement with people with whom one shares the present. "I can disregard, and dislike for their invasions on my reverie, the people in the streets who happen to compose the age in which I live." It is this fundamental disregard for people, and the irresponsibility produced by this indifference, that creates the conditions out of which West's reflections on the Nuremberg Trial are born. For if the Nuremberg Trial was conceived as a monumental instance of justice, a legal and historical edifice in line with London's impressive architecture, West's appraisal of this monument would offer a way to wander back into the past of the trial without the reassurance of comfort and with scarce attention to grandeur.

The Banality of Judgment

Rebecca West's *Daily Telegraph* pieces on the Nuremberg Trial and postwar Germany, a three-part series entitled "Greenhouse with Cyclamens"—a title whose innocuousness I address later in this chapter—are anything but stories of suspense. Indeed, West begins her report by undoing immediately any illusion that the story of Nuremberg will offer the same engaging appeal as a work of fiction, recounting her arrival at the Court of Justice in a tone bound to surprise her readers:

> It took not many minutes to get to the courtroom where the world's enemy was being tried for his sins; but immediately those sins were forgotten in wonder at a conflict which was going on in that court, though it had nothing to do with the indictments considered by it. The trial was then in its eleventh month, and the courtroom was a citadel of boredom. Every person within its walk was in the grip of extreme tedium.[15]

West lampoons the penchant for drama in these opening lines, building anticipation and deflating it just as quickly. Rather than approaching the trial as a watershed moment in international justice, she describes it as a staggering instance of boredom "on a huge historic scale" (*TP*, p. 11). The trial, simply put, had become excruciating for everyone: The judges, lawyers and secretaries, translators, and guards—all, that is, save the defendants—wanted nothing more than to have done with the entire affair.

> All these people wanted to leave Nuremberg as urgently as a dental patient enduring the drill wants to up and leave the chair; and they would have had as much difficulty as the dental patient in explaining the cause of that urgency. Modern drills do not inflict real pain, only discomfort. But all the same the patients on whom they are used feel that they will go mad if the grinding does not stop. (*TP*, pp. 7–8)

"The symbol of Nuremberg," West concluded pointedly, "was a yawn" (*TP*, p. 9).

Of the possible reactions to the Nuremberg Trial, West's was not what readers might have expected, and she likely suspected as much. For in spite of the seemingly bottomless quantities of evidence presented before the Tribunal, the event itself promised a terse drama before a rapt international audience. It would become "the greatest trial in history," declared Norman Birkett, one of the British judges. "The historian of the future will look back to it with fascinated eyes. It will have a glamour, an intensity, an ever-present sense of tragedy that will enthrall the mind engaged upon its consideration."[16] Robert Jackson, the chief American prosecutor at Nuremberg, reinforced this promise of juridical rapture in his powerful opening address. "That four great nations, flushed with victory and stung with injury stay the hand of vengeance and voluntarily submit their captive enemies to the judgment of the law is one of the most significant tributes that Power has ever paid to Reason,"[17] Jackson stated.

Yet by drawing out Nuremberg's boredom in a manner unsurpassed by her contemporaries—not to mention the historians, political and legal theorists, or philosophers who would write about it years later—West reminds us of a critical feature of the trial that turned out to be neither glamorous nor intense. Her observations in the Palace of Justice point toward a fundamental feature of law's encounter with catastrophe: that the legal process makes catastrophe available through ritualized but ultimately boring discourse. Rather than a flatfooted reflection, however, West's protracted meditation—her homage—to boredom raises vital questions about the relationship between being bored and doing justice. How can one understand an event that had promised to make the world breathless with suspense but that instead lulled it into a bored, indifferent numbness? What does this boredom do to, and for, West's analysis of the trial? More broadly, what does it do to law and to the experience of justice at Nuremberg? I intend to examine West's insistence on boredom not merely as a rhetorical attempt at irony, humor, common sense, or the sort of inappropriateness that produces shock value. Instead, I aim to explore it as a strategic and psychological process through which experience and, ultimately, memory are born.

"Greenhouse with Cyclamens" subtly but forcefully suggests that boredom constitutes not only a necessary feature of the law; it ultimately forms a critical dimension of the legal process by inviting the active creation of memory. West's response to Nuremberg thus offers a twofold argument: first, that trials will most likely be boring—that their structural features, from the recitation of evidence and their adherence to regulation, are bound to underwhelm. Beyond this fact of boredom, however, is West's second, and underlying, suggestion: that this tedium is productive, creating a space for memory that law, as a strictly procedural rather than a social practice, threatens to close off. This memory, in other words, becomes possible both because *and* in spite of the judicial process.

In his classic essay on the subject, "On the Psychology of Boredom," Otto Fenichel offers a valuable, if common-sensical, understanding of boredom as a state of mind that arises when *something expected does not occur*.[18] And indeed, the heightened expectations surrounding the Nuremberg Trial posed one of the most difficult challenges to those journalists charged with conveying its proceedings to a watchful world. How does one explain—and more crucially, make compelling—the fact that "something expected does not occur"? How, furthermore, does one remain faithful to these unmet expectations and still make the trial's experience meaningful?

Laurel Leff has noted that *The New York Times* reporters typically took the position of the prosecution, following the story from its vantage point with little attempt at objectivity.[19] West observed something else in the dominant newspaper accounts, which tended to focus on the trial's moments of drama, "concentrat[ing] on the sensational moments when the defendants cheeked back authority" (*TP*, p. 32). Her sensibility in "The Dead Hand" contained the kernel of what would become her skepticism of these accounts, which she saw as privileging outstanding, "monumental" moments bearing little relevance to the actual circumstances. They are dramatic, but "blankly not contemporary," and shed but little light on the trial as a whole.

The newspaper reports, however, did what one would expect from good (or at least savvy) journalism, presenting the trial in its exemplariness and thus informing as well as attracting readers. Along these lines, the court itself was often distinguished as an enclave of hope and possibility in contrast to its bombed-out, desolate surroundings. Janet Flanner, who reported on the trial for *The New Yorker*, saw precisely this contrast in the proceedings, and her remarks—as those of another highly regarded female journalist—offer an instructive contrast to West's account:

> The reason for the presence of us all here—an exotic, shut-off, quadripartite community, about two thousand strong—is the Tribunal, which seems a small island of hope, sanity, and justice surrounded by the sullen, Valhalla-minded Germans and their ruined town. There are a hundred and sixty journalists in Nuremberg—momentarily the world's largest news group in one place covering one event [. . .] We ourselves have only one daily newspaper to read [. . .] Phone calls to London are difficult, since they must go through a military switchboard to Frankfurt, be transferred there, through another military switchboard, to England, and then through another military switchboard to the London civilian circuit, while snowstorms have broken down the line in France. [. . .] In Nuremberg, all we know is what we see and hear.[20]

Flanner's "Letters from Nuremberg" privilege the court as a kind of consecrated space, even as they hint at the claustrophobia of its ad hoc community—a small price to pay for hope, sanity, and justice. There is a sense in Flanner's description, moreover, that Nuremberg's insularity oddly bolstered the trial's drama, setting the stage for a struggle of hope in the face of destitution. It is perhaps in view of this tension that Flanner, even as she documented the trial's atmosphere, devoted her reflections primarily to judicial issues in various courtroom scenes. Her reports on the trial never veer far from the law, staying close to its proceedings and probing their implications with acumen and focus.

Flanner's juxtaposition of courtroom and town extends the moral position adopted by the prosecution at Nuremberg, which invoked this contrast in framing the Tribunal's aim to restore civilization after barbarism. As Lawrence Douglas maintains,

> [O]ne important, if not inevitable, consequence of defining crimes against humanity as violations of civilized practice was that the prosecution came to characterize Nazi atrocities as *crimes of atavism*, horrific deeds committed in an orgy of primitive barbarism. As the Allies were paragons of civilization, the Nazis were consistently represented as atavistic.[21]

For West, however, the trial's contrasts—whether between court and town or humanity and barbarism—told little of what actually went on in Nuremberg. Unlike Flanner's "Letters from Nuremberg," "Greenhouse with Cyclamens" rarely delves into the legal substance of the case, straying far beyond the juridical to depict people and places in the town. Where Flanner found "an exotic, shut-off, quadripartite community" in Nuremberg's enclave of journalists, for whom the Tribunal represented a privileged, if hermetic community—"a small island of hope, sanity, and justice"—West identified a situation that bore an uneasy resemblance to the imprisonment of the men in the dock:

> What irked was the isolation in a small area, *cut off from normal life* by the barbed wire of army regulations; the perpetual confrontation with the dreary details of an ugly chapter in history which the surrounding rubble seems to prove to have been torn out of the book and to require no further discussion; the continued enslavement to the war machine. To live in Nuremberg was in itself physical captivity, even for the conquerors. (*TP*, p. 9, emphasis added)

West's "cut off" community is vastly different from the "shut-off" one that Flanner described. For where Flanner saw hope burdened with the inconvenience of seclusion, West sensed a troubling state of solitary confinement. The trial's isolation from its environs created a situation in which history was dealt with in a vacuum, becoming the purview of the Tribunal but bearing little relevance—and consequently generating little discussion—for those not directly charged with carrying out justice. Like the buildings of London that one passes "in the drooping of an eyelid," the trial threatened to become an event that one can revisit in the past tense, without consideration for its relevance to everyday life beyond it.

Instead of approaching the trial as a study in contrasts—noting rifts between the dignified Tribunal and the destitute city and seeing them as personifications of virtue and evil—West saw the trial and the town as parts of a larger whole. In advancing this claim, "Greenhouse with Cyclamens," as the mundane reference in the title aptly suggests, set the trial side by side with the strange if quotidian life that surrounded it: Rather than intruding upon everyday life, it seemed to blend—surprisingly and disturbingly—with its contours. "The trouble with Nuremberg," West writes, "was that it was so manifestly a part of life as it is lived. The trial was of a piece with the odd things that happened on its periphery, and these were odd enough" (TP, p. 55).

West is drawn to this periphery because it helps explain a great deal about the jarring encounters one experienced not only in Nuremberg but also in Berlin and presumably throughout much of postwar Germany. For just as the Nuremberg Trial surprised in its dullness, Berlin in 1946 caught one off guard by presenting all manner of seemingly "normal" encounters that, were one to scratch the surface, would reveal themselves as nothing more than façades. Thus, a bookstore that holds out the promise of a leisurely read turns out to contain little save Allied propaganda, and a bustling café, when one enters it, is full of people perched gloomily over cups of watery coffee. The irreconcilability of appearance and reality meant that "one was always being disconcerted by coming on a familiar form without its familiar content" (TP, p. 35).

This discrepancy between form and content turned out to be one of the most perplexing aspects of Nuremberg, precisely because the form one had expected—the unfolding of a gripping legal drama—buckled under the weight of the trial's "citadel of boredom." And if the drama did not unfold as expected, West suggests, this was partly because the narrative so often accorded it, the story of contrasts, had little to do with the actual relationship between courtroom and town.[22] The task of amplifying this relationship took West outside the court's confines and into the outskirts of the town that housed it. In her wanderings, she found not Janet

Flanner's "sullen, Valhalla-minded Germans" but an assortment of strange and curious individuals who defied expectation and who consequently could not be summed up handily, and certainly not in legal terms.

West's juxtaposition implies, moreover, that the form–content opposition exists as part of the trappings of law itself—that the law takes extraordinary events and people and concisely sums them up. By describing simple individuals who defy simple explanation, her narrative adds force to the idea that Nuremberg was not idiosyncratically boring but that there was something structural and essential about its boredom. Put simply, this "flattening out" is the way trials operate; that is, to do justice *in* law is not to do justice *to* life. And this is as it should be, for in filling in the picture of Nuremberg's surroundings, emphasizing those extralegal encounters that contribute to the overall experience of the trial, "Greenhouse with Cyclamens" illustrates the importance of creating memory independently from, but simultaneously with, law. It is not enough to assume that law, strictly conceived, will do the work of memory or even set the terms by which the past is remembered. To assume as much would be to relieve those uninvolved with the trial of the burden of remembering it or the catastrophe it sought to address. Something else—some other narrative—needed to be embedded in the legal story to make the trial memorable. West's writing, I propose, fashions this other narrative and, with it, the terms of memory set by the peculiar encounters in Nuremberg's periphery. If, as Otto Fenichel noted, boredom arises when "something expected does not occur," Rebecca West comes at this formulation differently, shifting the focus to the unexpected and inviting the question: If the expected does not occur, what does?

Her answer appears in the form of a series of encounters in Nuremberg's environs, the first of which involves an old woman whom West and her companions met in a village just outside Nuremberg on the evening before the trial's last day. The woman popped her "frizzled and grizzled head" (*TP*, p. 55) over a fence and proudly announced to the group that she "shot their King Edward." When they finally understood her to mean that she had gone shooting with King Edward, the woman launched into a tirade about the men on trial, inquiring whether Sauckel had been convicted. She sincerely hoped he had been because she held him personally responsible for "bringing these wretched foreign labourers into our Germany." When someone murmured that these workers would rather have stayed home, the woman heartily agreed: "Yes, yes, of course they should have been left at home, the place for a pig is in the stye. Oh, hanging will be too good for Sauckel, I could kill him with my own hands" (*TP*, p. 56). In her eagerness to speak with the English visitors, she then asked why, although "of course it was terrible what Hitler did to the Jews," the British nonetheless saw fit to appoint a Jew as their chief prosecutor. When someone meekly informed her that Sir David Maxwell-Fyfe was not a Jew, she laughingly admonished, "Oh, you English are so simple; it is because you are aristocrats. A man who calls his son David might tell you that he was English or Scotch or Welsh, because he would know that you would believe him. But we Germans understand a little better about such things, and he would not dare to pretend to us that he was not a Jew" (*TP*, p. 57).

West cuts from this episode to a stream a few miles from the village, where she happened upon a man doing his exercises. He was thin but muscular, she remarks, and one could tell that his leanness was the result of exercise rather than starvation. After describing the scene in careful detail, she concludes that it held no deep meaning about the differences between the Germans and the English.

> [T]here was no threat of perversity about his tranced and anxious stare. Simply he had come a long way to bathe in a stream which here was clean, as it would not be when it reached the fouled city where he had to live, and now he was affirming that though he had lost everything else he still had his body, he still had that surely quite remarkable stomach muscle, he still was his unique self. (*TP*, p. 58)

On the heels of this interlude, West describes a woman who worked as a servant in the villa that housed the trial's guests. Passing West's group on the road to the village, she stopped to ask whether Streicher had been sentenced. Upon learning that a sentence had not yet been passed, the woman grew silent for a moment and then declared that she hoped he would be hanged. She had been hoping for his punishment even before the war, when Streicher had visited the village and she and her husband had taken their children to the Town Hall to hear him speak. The speech, she recalled, began innocuously enough by addressing politics. But Streicher then shifted into "filth, gibbering filth about the Jews, describing the sexual offences he pretended they committed and the shameful diseases he pretended they spread, using dreadful words" (*TP*, p. 59). When she and her husband tried to usher their children out of the hall, several SS men forced them back to their seats. West re-creates the moment, invoking the woman's shock and indignation: "Yes, quite young boys had forced her and her husband to stay with their children in this bath of mud. Again she fell silent, and her face was a solid white circle in the dusk. Then she burst into a rage of weeping, and went away" (*TP*, p. 59).

Each of these profoundly human, disturbing, or odd vignettes is embedded in West's portrait of Nuremberg without accompanying explanation. For how does one respond when sentiments—like those that Sauckel and Streicher should be convicted—make sense but when the reasons behind them seem to miss the larger picture? None of these people seemed to strike West as particularly evil, and yet there is something deeply unsettling, even insidious, about their skewed, myopic vision of the horrors of the recent past. It is as though the lens through which the trial is viewed had lost its focus, leading one into conversations with individuals who believed that, indeed, the Nazis should be punished, who were outraged at what happened to the Jews, and yet who had come to these convictions from the unlikeliest of vantages. Something expected, indeed, did not occur, but that something had nothing to do with what was happening inside the Palace of Justice.

In her refusal to speculate about the meaning of these episodes, West seems to undercut the Tribunal's pedagogical aim. While the prosecution worked to make a sea of documents intelligible, and thus to make catastrophe understandable, her depictions of life outside the Palace of Justice upset this clarity and conveyed the experience of Nuremberg as a series of muddled encounters that resist being streamlined into tidy explanations. Yet this muddle, West implies, was part of daily life for those who sat in the Palace of Justice and waited for the defendants to be sentenced. The trial's importance as a landmark of international justice, a status with which West certainly agrees, ultimately had no bearing on the hearts and minds of ordinary people, who continued, in ways both jarring and ludicrous, to applaud the Tribunal's convictions for their own petty reasons.

Rather than serving as glosses on the trial, these strange encounters extended its quotidian dimensions and particularly what West saw as the ordinariness of the men in the dock. For what she means in claiming that "the trial was of a piece with the odd things that happened on its periphery" is that the strange, noteworthy, or downright startling was perpetually enmeshed with the most banal of details. And so it was that the accused looked nothing like the terrifying criminals one had expected. These expectations might be surmised from Janet Flanner's description of Goering on the stand: "There were not more than a few hundred people in the courtroom. They had the awful privilege of listening to the personal recital of a man who helped tear apart millions of lives, as if with those large, white hands that gestured as he sat on the Nuremberg witness stand."[23]

For West, such moments of "awful privilege" were scarce indeed. Instead, those who witnessed the trial found themselves straining to discover traces of evil in the defendants' resounding ordinariness: "Not the slightest trace of their power and their glory remained," West writes; "none of them looked as if he could ever have exercised any valid authority" (*TP*, p. 4). The trial made it difficult to distinguish the defendants as personifications of evil: "So diminished were their personalities that it was hard to keep in mind which was which, even after one had sat and looked at them for days; and those who stood out defined themselves by oddity rather than character" (*TP*, pp. 4–5). Goering and his large hands did not produce terror in the hearts of spectators; according to West, he struck a far more risible pose: "Goering still used imperial gestures, but they were so vulgar that they did not suggest that he had really filled any great position, it merely seemed probable that in certain bars the frequenters had called him by some such nickname as 'The Emperor'" (*TP*, p. 4).

Similarly, other defendants were far from memorable or even, for that matter, noticeable. Even those who "were still individuals" (*TP*, p. 5) looked strikingly like people one might find in the streets of Nuremberg or anywhere else. Streicher, West muses,

> was a dirty old man of the sort that gives trouble in parks, and a sane Germany would have sent him to an asylum long before. Baldur von Schirach, the Youth Leader, startled because he was like a woman in a way not common among men who looked like women. It was as if a neat and mousy governess sat there, not pretty, but never with a hair out of place, and always to be trusted never to intrude when there were visitors: as it might be Jane Eyre. And though one had read surprising news of Goering for years, he still surprised. He was so very soft. Sometimes he wore a German air-force uniform, and sometimes a light beach suit in the worst of playful taste, and both hung loosely on him, giving him an air of pregnancy. (*TP*, pp. 5–6)

Nor did this flatness of character change once the defendants had been convicted. Thus, when Goering made his final appearance in court, "it was not evident that he was among the most evil of human beings that have ever been born" (*TP*, p. 63). If this claim sounds at all familiar, perhaps this is because it would be made nearly two decades later, with the 1963 publication of Hannah Arendt's *Eichmann in Jerusalem*. Indeed, the connection between Arendt's famous (and famously controversial) *Report on the Banality of Evil* and West's "Greenhouse with Cyclamens" has yet to be fully acknowledged or addressed, but I would suggest that Arendt's position owes much to the one set out in "Greenhouse with Cyclamens." West, however, offers more than a reflection of banal evil: She actively fashions this banality herself, giving us some of the most unlikely descriptions of the men in the dock. And in doing so, she emphasizes the importance of the trial as a first-order experience rather than a basis for a wider philosophical claim about evil—a focus that may go a long way toward explaining why her reflections have been eclipsed by Arendt's.

In focusing on the banality of Nuremberg, West suggests that the difference between what is remembered and what is forgotten, between what keeps us in suspense and what causes our attention to slacken, hinges on the difference between what is distinct and what is not. This is implied in West's own turn of phrase: an issue of whether something is "*a part* of life as it is lived"—integrated with it, indistinguishable from it—or whether it stands *apart* from it, differentiating itself in a way that lops off past from present, good from evil. These distinctions, West insists, were manifestly unavailable in the Nuremberg Trial.

What distinguished the trial was, ironically, a *refusal* to distinguish, in certain crucial instances, between those who were being judged and those who sat in judgment. West thus notes the Tribunal's acquittal of German naval admirals on the grounds that Allied forces had also engaged in unrestricted submarine warfare. "This *nostra culpa* of the conquerors," she reflects, "might well be considered the most important thing that happened at Nuremberg. But it evoked little response at the time, and it has been forgotten" (*TP*, p. 53). It has been forgotten, West implies, in part because it is plainly unpalatable. But it has also been forgotten because it did not make the accused stand out, instead eroding the distinction between prosecution and defense and undoing the "rhetoric of atavism"[24] that Lawrence Douglas identifies as a crucial part of Nuremberg's grammar. And yet this blurring of categories conveys something fundamental about the difficulty of sitting, day after day, in a trial whose beginning had become a distant memory and whose end was nowhere in sight.

The Trouble with History

The story that "Greenhouse with Cyclamens" tells is one in which law—and West herself—flattens the robust stories it had been expected to produce. But it is *this* story, rather than our expectations of Nuremberg, that conveys the trial *in all its contemporaneousness.* This does not mean, of course, that West's report is free from the very kinds of judgments and conclusions that a historian might reach. But for the historians, the Nuremberg Trial would be consigned to the past, judged in relation to events that preceded and followed it. Whether the trial bored or excited its audience is secondary to whether it passed judgment with all deliberate speed and created a legal infrastructure to respond to unprecedented atrocities. Moreover, the priority given to this infrastructure, and to the concomitant dullness it produced, was not unique to Nuremberg; it is simply what trials do.

West's reflections on the difference between lawyers and historians suggest as much, insisting in broad terms that the historian's insights are no more valuable than the immediacy of an imperfect legal process. For what is lost in the historian's slow, measured path through the trial's immense body of documents is, put simply, a clear sense of the present. Faced with this overwhelming accumulation of evidence, West writes,

> the lawyers had to act with a haste that was all too good. The historians would have taken them to their studies, shut the doors, and dealt with them at the slow pace of scholarship, and scholarly prejudices and obsessions would have struck deep roots and grown a quickset hedge about the facts before the work was done. The glands of some don might have a second springtime, or even a first, while he weaved an erotic fantasy about Goering; and a don of the other sort . . . might find it in a thesis that Goebbels was a man of elevated character and who might have saved Europe had it not been for the machinations of the British Foreign Office. (*TP*, pp. 261–62)[25]

Historians would have done what scholars, by nature or profession, are in the business of doing: distinguish themselves. And this distinction, while it might make for more gripping reading than the tedious record of the Nuremberg proceedings, would impede the pursuit of justice or, worse still, generate a far more flawed assessment of the past than any produced by the hastily convened Tribunal.

The primacy West ascribes to lawyers over historians is more vociferous than the prevailing sentiments so often expressed about Nuremberg. The prosecution undoubtedly saw the need for law to act swiftly so that historians could work through its findings slowly and carefully, and few indeed cast aspersions on the historian's role. From Norman Birkett's pronouncement that "The historian of the future will look back to it with fascinated eyes" to chief prosecutor Robert Jackson's conviction that "its full development we shall be obliged to leave to historians,"[26] there is a sense of hope, relief, and confidence that the trial would (and should) be left to history.

But West was unconcerned with history in any professional or pedagogical sense. Instead of a clear assessment of the trial—a legal or political parsing that might be of use to historians—she recounted a seemingly haphazard, unfocused record. What was she after in turning away from the historian's position? What sorts of distinctions matter for West? That is, in whose voice is her account written?

If there was a distinction worth making in Nuremberg, it was between those who were there to witness the trial and those who were not. For if "the trouble with Nuremberg is that it was so manifestly a part as life as it is lived," this difficulty was not limited to historians or to those who were present in the courtroom. Above all, it troubled those who read about it from a distance, turning to the newspapers to get a sense of the events in Germany. And it is to these people that West addressed her account of the trial, not simply as a reporter but above all as a witness.

> It was necessary, and really necessary, that a large number of persons, including the heads of the armed and civil services, should go to Nuremberg and hear the reading of the judgment, because in no other conceivable way could they gather what the trial had been about. Long, long ago, the minds of all busy people outside the enclave of Nuremberg had lost touch with the proceedings. The newspaper reports inevitably concentrated on the sensational moments when the defendants cheeked back authority, and to follow the faint obtrusions of the serious legal issues which made their way into the more serious journals would have taken the kind of mind which reads its daily Scripture portion and never misses; and that kind of persistence carries one irresistibly to the top of the grocery store, and no further. (*TP*, pp. 32–33)[27]

The necessity of witnessing the trial, and the blindness of those who did not witness it, drives West's report. For "the trouble with Nuremberg," it turned out, was that those who heard about it from afar received a false sense of what took place on a daily basis and thus would remember the trial only through its dramatic representations in the public press.[28] But these exceptional moments, like the strict legal assessments of the Tribunal, did not convey "what the trial had been about." For those who sat through its eleven months, the trial's states of exception—those moments that would be showcased in the newspapers and preserved in historical records—only underscored the experience of being "cut off from normal life." Reaching beyond this isolated enclave and blurring the distinctions between life inside and outside the courtroom, West bore witness to Nuremberg as "a place of sacrifice, of boredom, of headache, of homesickness" (*TP*, p. 18).

In framing her essays not just as journalistic accounts but also as eyewitness testimonies, Rebecca West infused the trial with one of the very components it lacked. For the prosecution had made the calculated decision to downplay human eyewitnesses in favor of documentary evidence, responding to the fear that individuals on the stand would be far less reliable and far more impeachable than captured Nazi documents. "We must establish incredible events by credible evidence," Robert Jackson wrote in a report to President Truman the summer before the trial.[29] West might thus be seen as supplying the human voice—not from the witness stand but from the press gallery.

The Possibility of Memory

West's problem with historians takes root in her worry that in being subject to history, the trial will no longer be available to *memory*: The historian, or the lawyer, might make diligent use of its reams of documentation, but the event will not seep back into collective memory. Similarly, even as its juridical legacy created a precedent through the legal category of Crimes Against Humanity, the memory of what it felt like to be part of this moment of justice would elude even the most thorough historian. Like London's distinguished architecture, the trial would acquire historical grandeur and in the process would become "blankly not contemporary," set apart from everyday concerns, much like the trial itself.

The prospect of forgetting clearly troubled West, accounting for her refusal to lapse into terse drama or legalese in her report of the trial. Before long, she reasoned, everyone would return home, and Nuremberg would become a distant memory. "The trial had begun to retreat into the past. Soon none of us, we believed, would ever think of it, save when we dreamed about it or read about it in books" (*TP*, p. 70). These dreams, however, will never quite pierce the fabric of daily existence; instead, they threaten to relegate the trial, and the catastrophe to which it responded, to the realm of the unreal.

West's concern over events that are assigned to history but forgotten by the present helps explain her emphasis on banal details and everyday stories. For in casually relaying the events at Nuremberg as though she were speaking to someone over dinner rather than writing for the public press, West created an enduring conversation. That is, she sought to tell stories about Nuremberg that bear *repeating rather than recording*. To record something relegates it to the annals of history, transporting it out of life's bustle and into a locked study—a move that West will not countenance.

Rebecca West's approach to the trial does more than combine what her first biographer describes as "hard investigative reportage with theatrical scene-setting"[30]—a view of West's work that sets it in line with countless other examples of colorful journalism, as well as the intellectual view of law as a public drama, a "theater of justice." In defiance of these positions, she shaped a response to the Nuremberg Trial's "boredom on a huge historic scale," a mnemonic technique to absorb its enormity while skirting its boredom. She undertook, in other words, not only the task of reporting the present but also of creating a complex map of memory. In supplying the trial with a context, giving it a sense of place, West's report suggests that memory demands a location, a *lieu de mémoire*. This phrase, borrowed from historian Pierre Nora,[31] calls to mind the cultural and national phenomenon of "housing" collective memory. "Greenhouse with Cyclamens," however, recasts this sense of memory. Rather than anchoring it in one particular location, most prominently the Palace of Justice, West lays out an unanchored series of places, a collection of sites that bear no resemblance to monuments, memorials, or other consecrated "spaces of memory."

In carving out a landscape of people and places, West's peripatetic approach to the trial reaches beyond Nora's notion of collective memory. Her sense that place is vital to the act of remembering takes root in the ancient and medieval elaboration of the architectural mnemonic, whereby an individual maps the details to be remembered onto the backdrop of a physical location. "The structure of memory, like a wax tablet, employs places [loci] and in these gathers together images and letters,"[32] argued Cicero. But it is not enough, in West's account of Nuremberg, to think of places to remember the legal or historical import of the trial. One needs to encounter these places and people fully, engaging with rather than just employing them. The act of remembering thus becomes, at this unprecedented historical juncture, a lived *experience* rather than an instance of recall.

In keeping with this notion, "Greenhouse with Cyclamens" lays out a mnemonic technique not unlike that used by individuals with remarkable memories. In drawing upon the technique of using places as memory prompts, West's wanderings through Nuremberg resemble those of one of the most famous case histories of remarkable memory, which was recorded by A. R. Luria in *The Mind of a Mnemonist*. I would like to draw upon the extraordinary memory of Luria's patient to articulate more concretely how West shapes her reflections on Nuremberg.

Mnemonic Wanderings

In the mid-1920s, Soviet psychologist A. R. Luria was introduced to a man who would become the most challenging patient of his clinical career. This remarkable individual, Sherashevsky—whom Luria referred to in his writing as S.—appeared to have an inexhaustible memory. Luria followed his subject over a period of thirty years, during which he tested him by giving him series of words, numbers, or letters, which S. could reproduce in any order and which, even after the passage of several years, he did not forget. But rather than putting his subject to the test, the study called Luria's own work into question. "[I]t appeared that there was no limit either to the *capacity* of S.'s memory or to the *durability of the traces retained.* Experiments indicated that he had no difficulty reproducing any lengthy series of words whatever, even though these had originally been presented to him a week, a month, a year, or even many years earlier."[33] Memory, it seems, had called scientific method into question; the capacity to measure and take note had broken down in its overwhelming presence. Rather than measuring S.'s memory, Luria can only describe it:

> When S. read through a long series of words, each word would elicit a graphic image. And since the series was fairly long, he had to find some way of distributing these images of his in a mental row or sequence. Most often [. . .] he would "distribute" them along some roadway or street he visualized in his mind. Sometimes this was a street in his home town, which would also include the yard attached to the house he had lived in as a child and which he recalled vividly. On the other hand, he might also select a street in Moscow. Frequently he would take a mental walk along that street—Gorky Street in Moscow—beginning at Mayakovsky Square, and slowly make his way down, "distributing" his images at houses, gates, and store windows. At times, without realizing how it had happened, he would suddenly find himself back in his home town (Torzhok), where he would wind up his trip in the house he had lived in as a child.[34]

S.'s strategy for remembering integrates the strange with the familiar, embedding an unrelated series of words within a topography that he can imagine, and revisit, in his mind's eye.[35] S. described his infallible memory not as the result of focus, of the sort that one might summon by staring intently at an object so as to recall it later. Instead, he explained, "I recognize a word not only by the images it evokes but by a whole complex of feelings that image arouses. It's hard to express . . . it's not a matter of vision or hearing but some over-all sense I get."[36] The images that S. sees, moreover, do not function as symbols, each serving as a lightning rod to which he can then "append" the words he needs to recall. Instead, they serve as a kind of map that, when animated in his mind, he can follow by intuition rather than by rote.

It is this "over-all sense," I would suggest, that West pursues in narrating the Nuremberg Trial through its ordinariness. Her approach does not make the strange familiar. She does not seek to make her readers fluent in the language of law but instead integrates the trial, in all of its strangeness, with ordinary encounters, with the odd individuals she encounters outside the courtroom. These resolutely accessible moments chart avenues of experience through which one can trace one's way not to the Tribunal's juridical signposts but to the broader, more unwieldy "complex of feelings" of "what the trial had been about." Rather than pinning down Nuremberg to a legal turning point or articulating it with the resounding voice of judgment, she gently coaxes an entire world out of the courtroom, a world as perplexing as it is real.

The infinite capacity of S.'s memory, however, turned out to be both a blessing and a curse. For his ability to remember everything also meant that he forgot nothing, and was thus incapable of making distinctions, of separating out the things that matter from his vast chain of memories. In his Foreword to Luria's case history, Jerome Bruner writes that the psychologist's account of S. sheds light on "what it means for somebody to live with a mind that records meticulously the details of experience without being able, so to speak, to extract from the record what it means, 'what it's all about.'"[37] The nonselectivity of S.'s memory means that "what remains behind is a kind of junk heap of impressions."[38] Reading a simple passage in a book thus becomes a painfully complicated procedure for S. because every word calls up an image that prevents him from grasping the overall meaning of a sentence. As he described it,

> I'm slowed down, my attention is distracted, and I can't get the important ideas in a passage. Even when I read about circumstances that are entirely new to me, if there happens to be a description, say, of a staircase, it turns out to be one in a house I once lived in. I start to follow it and lose the gist of what I'm reading.[39]

If his description feels very much like West's digressions in "Greenhouse with Cyclamens," perhaps this should not surprise us. For her, however, this inability to distinguish, to have one image dissolve into the next, was an intrinsic part of what it felt like to be at Nuremberg. It was the feeling of being confronted with an event that threatened to collapse into just such a "junk heap of impressions": a confusing, overwhelming accumulation of evidence spread over a seemingly interminable progression of days, none of which resolved themselves into any obvious schema. Yet this muddle of one moment shading into the next suggests what it felt like to be at the trial, without history books or hindsight to confer meaning and perspective. And whether one sees this junk heap as sensory overload or sensory deprivation—the result of being asked to recall arid details like S.'s numbers or the trial's numerous legal documents—one can readily acknowledge the urgency of finding meaning in the overload created by the trial. For what the observer remembers from Nuremberg will lay the foundation for what future generations recall and whether this past matters, not to historians or philosophers, but to ordinary individuals for whom the past is not a vocation.

The pressure that West felt to commit the trial to memory, and the mnemonic wanderings that emerged from this commitment, illuminate a crucial difference between the particular demands of memory after each world war. For memory after the Great War was seen in the light of its widespread instances of shell shock, and so the duty to remember was accompanied by the need to recover memory. But if recovery was the focus of West's narrative tension in *The Return of the Soldier*, this focus was eclipsed by her determination to *build up* memory at Nuremberg, fashioning it out of the trial's bottomless tedium. Faced with the burden of forgetting, and deprived of the kind of experience that would be interesting enough to compel memory in its own right, West set out to approach this experience through other routes. There was, in short, nothing to recover, and so memory had to be fashioned out of the homelier episodes that kept boredom at bay.

From Event to Experience

The boredom that initially appeared in "Greenhouse with Cyclamens" as part of the story of Nuremberg's unmet expectations—its lack of distinctions and odd interludes—also gave shape to still another story about the trial. For in spite of its slow pace, which carried the weight of deliberation and a juridical steady hand, the Nuremberg Trial would never be able to fully absorb the scope of the catastrophe it was designed to address. Perhaps this is the nature of law generally, just as it was part of Nuremberg particularly. As West observed one year earlier at the 1945 trial of William Joyce, who was tried and convicted in London for broadcasting Nazi propaganda in England during the war:

> For the law, like art, is always vainly racing to catch up with experience. [. . .] By a gross inappropriateness judges and legislators are described always as sitting, the one in the bench, the other in the House of Commons or the House of Lords; for in fact they run, they run fast as the hands of the clock, reaching out to the present with one hand, that they may knot it to the past which they carry in their other hand.[40]

It is in this race toward an experience that can never be grasped fully that Nuremberg faltered as well. For all of its staggering documents, matched by its staggering dullness, the trial sought an experience, and sought to *become* an experience, that was never attained. The trial, West reflected, was "an unshapely event, a defective composition, stamping no clear image on the mind of the people it had been designed to impress. It was one of the events which do not become an experience" (*TP*, pp. 262–63).[41]

West's conception of law pits the trial's tediousness against the law's breakneck speed—a tension that epitomized Nuremberg at its deepest level. "Here was a paradox," she writes. "In the courtroom these lawyers had to think day after day at the speed of whirling dervishes, yet were living slowly as snails, because of the boredom that pervaded all Nuremberg, and was at its thickest in the Palace of Justice" (*TP*, p. 18). Caught between this swiftness and its contrapuntal tedium, it became impossible to grasp, and thus to absorb, the encounter with justice in any enduring way. West's essays offer their most powerful intervention in this race to catch up with experience: By drawing out the trial's boredom, they clear a temporal space, slowing down the pace of justice and making discovery—and the experience in its wake—possible.[42]

In narrating the trial through boredom's slowness, moreover, West's essays resist the pressure that made Nuremberg a nonexperience: the pressure to be interesting, to hold the audience's attention in the manner of a Greek tragedy. The insistence that one be interested, psychotherapist Adam Phillips maintains, begins early in life and prevents us from seeing just how productive boredom can be. "It is one of the most oppressive demands of adults that the child should be interested, rather than take time to find what interests him. Boredom is integral to the process of taking one's time."[43]

This commitment to taking one's time, to stepping out of the law's race toward experience, motivates West's insistence on dullness. Nuremberg's tedium became not simply a description of the trial but an essential part of her effort to make it available to memory as an experience in its own right. For in creating a mnemonic map that brought Nuremberg into the cultural imagination, West needed more than creative license: She needed license to wander, literally and figuratively, beyond the court's confines, discovering the people and places that made the trial "a part of life as it is lived."

Examining this life for memory's sake, "Greenhouse with Cyclamens" encourages a return to boredom not as absence, as a lack of drama or excitement, but as *presence*. For what we remember about Nuremberg through West's reports is not just that a trial took place; we learn, with great precision, about the steady, driving presence of dullness. The tedium of legal procedure may well be the price of making catastrophe manageable; in that sense, Nuremberg's boredom is frustrating. But its boredom is ultimately an example of what the law, in its failings, makes possible: a memory that can—indeed must—emerge through extralegal means. The fact that Nuremberg was uninspiring in its tedium may disappoint, but the notion that this tedium invites memory work offers reassurance that boredom can become, in spite of the law, a means to other (and quite surprising) ends.

West's perambulations and the unlikely vignettes of Nuremberg they yielded— stopping points where memory could take hold—bring her approach closer to an analytic than to a legal sensibility. Adam Phillips, in his reflections on psychoanalysis, helps uncover the implications of her analytic sensibility. Phillips explains the ethos behind psychoanalysis by drawing on Winnicott's experiments with children. After placing a silver tongue depressor on the table in front of a mother and her child, Winnicott would wait to see whether the child took hold of the shiny object on his own or whether he would be handed or in some way forced to interact with it. Phillips comments on the experiment:

> For the child to be allowed to have what Winnicott calls *"the full course of the experience"* the child needs the use of an environment that will suggest things without imposing them; not preempt the actuality of the child's desire by force-feeding, not distract the child by forcing the spatula into his mouth. It is a process, Winnicott is saying, that is easily violated—although I would say that in growing up one needs *a certain flair for distraction*—and analogous to the analytic situation, in which the analyst's interpretations *offer views rather than imposing convictions.*[44]

In a manner related to both Winnicott's and Phillips's understanding of boredom, West's attitude toward Nuremberg can be understood as more analytical process than legal reflection. Conviction—whether of a court, a parent, or an analyst—intrigues her far less than the disparate views and experiences that one discovers when one allows oneself to indulge in distractions. This permission to be sidetracked, to leave the courtroom and instead to stumble across all manner of people and things outside it, shapes West's response to the trial. For her, telling the story of how the law functions may explain how a legal mind works, but it tells very little about how a bored mind works, and it is *this* mind that ultimately shaped, and was shaped by, the experience of the trial.

In allowing Phillips's "flair for distraction" to dictate the course of her essays, West was untroubled by the lack of focus that so plagued Luria's S. For if her experience resembles that of the mnemonist—"I'm slowed down, my attention is distracted, and I can't get the important ideas in a passage"—this slowness was precisely what she needed to get at "the full course of experience." The important ideas that elude S. are, in her mind, the purview of lawyers and historians, whose job is to pare down the trial to its pivotal moments and key concepts; their analysis, however, will leave little to remember about what the confusing, dull experience of justice felt like.

If West arrived at this experience through the vicissitudes of boredom, she made it memorable by withholding interpretation. The unexplained encounters in Nuremberg continue to perplex, and as Walter Benjamin reminds us, it is this unanalyzed core that makes memory possible:

> There is nothing that commends a story to memory more effectively than that chaste compactness which precludes psychological analysis. And the more natural the process by which the storyteller forgoes psychological shading, the greater becomes the story's claim to a place in the memory of the listener, the more completely is it integrated into his own experience, the greater will be his inclination to repeat it to someone else some-day, sooner or later. This process of assimilation, which takes place in depth, requires a state of relaxation which is becoming rarer and rarer. If sleep is the apogee of physical relaxation, boredom is the apogee of mental relaxation. Boredom is the dream bird that hatches the egg of experience. A rustling in the leaves drives him away.[45]

A story is compelling, in Benjamin's understanding, because its meaning has not been exhausted; the moment it reaches its interpretive limit, it ceases to be told or repeated. There is nothing left to say, nothing worth (or in need of) remembering.[46]

This nearly natural impulse to explain something exhaustively, underscored in the image of the rustling leaves, must be overcome consciously. Benjamin and Phillips suggest that the inclination to explain is so powerful that avoiding it may feel like failure or at least a frustration of one's intellectual powers. And it is in this sense of frustration that we find an analytic connection between Rebecca West's reflections and A. R. Luria's case history. Faced with the inexhaustible scope of his patient's memory, Luria's capacity to explain—and as a researcher, to measure—failed him:

> As an experimenter, I soon found myself in a state verging on utter confusion. An in-crease in the length of a series led to no noticeable increase in difficulty for S., and I simply had to admit that the capacity of his memory *had no distinct limits;* that I had been unable to perform what one would think was the simplest task a psychologists can do: measure the capacity of an individual's memory.[47]

He thus abandoned the prospect of analyzing his subject's memory and instead committed himself to describing it as fully as possible. "All this meant that I had to alter my plan and concentrate less on any attempt to *measure* the man's memory than on some way to provide a *qualitative analysis* of it, to describe the *psychological aspects of its structure.*"[48]

If we think of Luria's initial goal of measurement as analogous to analyzing an event—both attempt to harness and make sense of the unwieldy—we begin to understand the sort of impulse that West overcame in her approach to Nuremberg. Faced with an experience of overwhelming proportions, she offered no interpretation, no clear-cut legal analysis. Instead, she embarked on a series of associations and encounters that described rather than explained, refusing to restrict the trial to a particular legal moment or to the unequivocal pronouncement of a verdict. When explanations fell short, when the trial confused more than it clarified, she turned instead to "the psychological aspects of its structure" and committed these aspects to memory.

Judgment's Context

The "flare for distraction" that overtook West as she covered the Nuremberg Trial led her to what arguably became the unlikeliest discovery of all, the centerpiece of her writing on the trial: the greenhouse with cyclamens. For if "the symbol of Nuremberg was a yawn," West supplanted this yawn with a resolutely ordinary image, yet one so compelling for her that she took it as the title for all three of her Nuremberg essays. In a trial that offered more than its share of potential symbols, from the Palace of Justice, to the prisoners in the dock, to the horrific, unforgettable images of the shrunken head of a hanged Pole and of Holocaust victims depicted in the concentration camp film,[49] West chose instead to present the trial through an image that seems altogether ill-fitting in its banality. "Often people said, 'You must have seen some very interesting sights when you went to the Nuremberg Trial.' Yes, indeed. There had been a man with one leg and a child of twelve, growing enormous cyclamens in a greenhouse" (*TP*, p. 127). What is the significance of West's focus on these cyclamens? In a report devoted to relating the experience of judgment and justice after the war, why did she not rely on symbols that would call up the trial directly, signaling Nuremberg's unique place in history and law in explicit terms?

West answered these questions through the relationship between boredom, distraction, and memory woven throughout "Greenhouse with Cyclamens." Bored and distracted, she noticed the greenhouse from the converted castle in which the journalists were billeted; intrigued, she set out to "see how the Germans had kept that form of luxury going" (*TP*, p. 28). Anticipating little more than one would find in an English greenhouse after the war—"a desert place of shabby and unpainted staging, meagrely set out with a diminished store of seed boxes" (*TP*, p. 28)—she found instead a place well beyond her expectations:

> But the door was open; and it admitted to a scene far distant in time and space. This might have been a greenhouse in one of the great English or Scottish nursery gardens before 1939; or one might push the date back further, to a time when labour was still cheap. There was perfect cleanliness and perfect neatness here, and it was full of plants [. . .] There was a row of canna lilies, scarlet and orange and crimson, bright with health; there were many obconica primulas, which perfectly exhibited their paradoxical character of being open-faced and brilliant yet recognisably members of a shy and cool family; and there were rows upon rows of beautifully grown cyclamen which would have done credit to a specialist firm. (*TP*, pp. 28–29)

This flourishing greenhouse, only a short distance from West's temporary residence in the converted Schloss, surprised her because it looked as though it belonged in another time, before the war had decimated any prospect of abundance and made luxuries like flowers a thing of the past. In an odd, unexpected moment, West stumbled upon an emblem of nostalgia, a scene that seemed as impossible in the present as it was comforting in the past.

But this was not the first time West had come across this image and taken it as emblematic of the experience of Nuremberg and postwar Germany. For flora had already established itself in many minds as of a piece with so many of the chilling ironies about the Holocaust, chief among them the irony that a nation of such refined culture could descend to unspeakable depths of barbarism. Shortly before her description of the greenhouse, West noticed flowers elsewhere, while speaking with the French doctor who maintained the exhibits of Nazi atrocities at the Palace of Justice. Turning in his hand a lampshade made of tattooed human skin, he reflected, "These people where I live send me in my breakfast tray strewn with pansies, arranged with exquisite taste. I have to remind myself that they belong to the same race that supplied me with my exhibits, the same race that tortured me month after month, year after year, at Malthausen [sic]" (*TP*, p. 23). The contrast between the gruesome object in his care and the delicate flowers he received each morning prompted West to contemplate not irony but mystery:

> And indeed flowers were the visible sign of that mystery, flowers that were not only lovely but beloved. In the window-boxes of the high-gabled houses the pink and purple petunias were bright like lamps. In the gardens of the cottages that bordered a road

which was no longer there, which was a torn trench, the phloxes shone white and clear pink and mauve as, under harsh heat, they will not do, unless they are well watered. It is tedious work, training clematis over low posts, so that its beauty does not stravaig up the walls but lies open under the eye; but on the edge of the town many gardeners grew it thus. The countryside beyond continued this protestation of innocence. A path might mount the hillside, [. . .] [but] it would not lead to any place where it was not plain that Germany was a beautiful country, inhabited by a people who loved all pleasant things and seemed to mean no harm. (*TP*, p. 23)

As the German countryside spread out before her in bright bursts of flowers, West remained deeply conscious of the darkness rent by their radiance. For it was not simply that these flowers made the postwar desolation a little more bearable, lining the bombed-out road with bright colors or restoring to the hillsides some of the loveliness they once possessed. Nor was the "mystery" of these flowers only a matter of irony born of resilience amid destruction or, more tragically, the irony of a nation of people whose "lovely and beloved" flora inspired an abiding affection that did not extend to human beings.

To be sure, West was attuned to these ironies but did not limit the flowers to this symbolism alone—almost as though she were resisting the temptation to make use of an all too convenient cliché. Instead, she immediately complicates this image of life amid death: Picking up on the lampshade in the French doctor's hands, she superimposed the artifact onto her reflections on the resurgence of plant life, carefully noting how "the pink and purple petunias were bright like lamps." In yoking the image of blooming life to the doctor's material reminder of atrocity, West cast aspersions on the "mystery" of these bright spots of life among the ruins. And in linking these two lamps—one a stark, all too literal reminder of death and the other a vestige of nostalgia or hope—West subtly suggested that one invariably begets the other. The light given off by the pink and purple petunias is brighter for having cast its beams through the thick darkness of the doctor's lampshade, and the very fact that one looked toward this light served as a bleak reminder of the past that so many in 1946 were anxious to forget.

In dwelling on the "visible sign of that mystery" of civilization and destruction, West veered briefly toward the sentimental or at least the existential. By making the floral lamp and human lampshade contingent images, drawing one out of the other, she backed away from any potential sentiment or philosophy. But her interest ultimately lay in more practical matters—in the intricacies of the life surrounding Nuremberg and of what, and who, made this life possible. And so she observed moments later how remarkable this spread of flora truly was given the postwar circumstances: "It might seem that it would never be very interesting that somebody had started a brisk business on potted plants. But this was Germany, this was 1946; and it was as if one were in a lock, and saw the little trickle of water between the gates which meant that the lock was opening. The war had burned trade off Germany as flame burns skin off a body" (*TP*, p. 29). And yet this trade, unlike the lampshade of human skin she described moments earlier, was capable of being revived and even, strangely enough, of resisting political restrictions. For it was likely that the greenhouse near the Schloss had been active during the war, flouting Hitler's regulations; it was just as likely, West speculated, that it was now defying Allied restrictions, which would be unlikely to approve of using German fuel and labor to grow flowers.

The greenhouse's existence was of a piece with West's broader interest in things set off from the life around them. In this sense, her presentation of the greenhouse invites comparison with another, equally isolated, and artificially constructed space, but one that emitted none of the life in which she now found herself: the concentration camp barrack. Indeed, the contrast between the two places could not be starker: one a moist, warm, life-giving space and the other a bleak vision of death and inhumanity at its most brutal. Her depiction of the greenhouse invoked its ghostly counterpart without explicitly comparing it, becoming another instance of the powerfully associative quality of West's memory.

This focus on the practical matters of the greenhouse's existence prompted West to consider the individual responsible for its upkeep. Contrary to her expectations, the greenhouse was not tended by several gardeners: "[T]here was only one man to be seen, who was closing a light at the other end of the greenhouse with a clumsiness which was explained when he stumped off to another light on two crutches. He had lost a leg" (*TP*, p. 30). West's observation initially seems a pathos-imbued description of an amputee who, in spite of his suffering, had found a means of livelihood in the bleak Nuremberg landscape—a testament to survival and determination. But her point is vastly different:

> The twitch and roll, twitch and roll of his walk, recalled another difference between the British and the German lot. The Nazi Government had shown a monstrous cruelty to its own people in two respects. They did not dig out their dead from the ruins after air-raids [. . .] Neither did they make the proper effort to furnish artificial limbs for their war casualties, and an appalling number of one-armed and one-legged men were to be seen in the German streets. (*TP*, p. 30)

The cyclamen grower, whose only assistant was a twelve-year-old child, thus became not just a testament to postwar resilience, nor an invitation to ponder the pity of war, extending such pity even to the citizens of the nation being judged at Nuremberg. Instead, he became emblematic of a more general tendency of his own government, which demonstrated brutality not just to those it deemed "outsiders" but exercised a chilling callousness even toward its "exemplary" citizens. As a casualty of war, this man was a particular instance of a widespread phenomenon and a reminder of a crucial difference between England and Germany that invited not only revulsion but above all judgment.

It is thus that West's complex description, which seems to have little to do with idioms of law or justice, steers its readers toward judgment nonetheless. The mind distracted by the proceedings inside the Palace of Justice seeks refuge in the steady revival of life outside its purview. And these extralegal experiences, West suggested, were also part of the experience of law, allowing one to pass judgment by other means, refusing to confine it solely to the official work of the Tribunal.

In creating a landscape of seepages, of justice leaking slowly and unpredictably beyond the framework created to contain it, West did not suggest the existence of two independent realms of judgment, one official and "artificial," the other organic and accessible in the ordinary world outside the court. For the cyclamen grower—the human image that invites at once curiosity, pity, and judgment— would not exist without the trial and the influx of potential customers it brought to Nuremberg. Indeed, his business began at the start of the trial, when he realized just how many foreigners would descend upon Nuremberg. Now he had come to depend on the steady stream of buyers and so, like the defendants themselves, he was not terribly anxious to see the Tribunal close its doors.

> He wanted to know how many trials were likely to be held in Nuremberg now that this one was finished, and whether as many Americans and British and French officials would be here to conduct these others; and it was plain that though he was aware that he would be told that the number would be less, he longed to hear that it would be not much less. He enquired whether any of the English people now here would be likely to stop off in Holland on the way home and would be able to send him Dutch seeds. He would have more to say, but the greenhouse was getting dark. (*TP*, p. 32)

West's story of the cyclamen grower is a narrative of contingencies: of a business brought to life with the onset of the trial, of a man she would not have encountered had the journalists' quarters not been sufficiently close to his greenhouse. Folded into this context, the cyclamens, in their surprising dependence on the trial and the catastrophe to which it responds, became delicate reminders of the life quietly unfolding around the site of judgment.

For the Nuremberg Trial not only restored justice after the war; it also brought to life a world around it—a life that could only be glimpsed when one left the Palace of Justice to explore its bombed-out surroundings. From this perspective, the trial set the tone for a future that its own architects and participants, because they were only temporary visitors, would never see. Justice, West thus implied, has consequences that far exceed verdict, sentence, or for that matter, law.

West's symbolism unfolds as a web of associations: It is always relational, reminding one of the complexity of symbolizing justice and history. The cyclamens, which appeared in her essays initially as one-dimensional symbols of beauty and civilization in an era of barbarism or of survival in the wake of destruction, became for her emblems that demanded a *context* and that stubbornly insisted that justice was not a solitary pursuit. Rather than offering the cyclamens as self-contained images, she embedded one symbol in another: the greenhouse *with* cyclamens, attended *by* a one-legged man, all of which are situated in the shadow of the Schloss, which is itself in the shadow of the Palace of Justice.

By invoking the cyclamens in her effort to describe and, subsequently, to remember Nuremberg, West's complex symbolism recalled another image from World War I: the Flanders poppy. As a figure of postwar memory, the poppy was claimed for the poetic imagination in one of the Great War's most famous poems, John McCrae's "In Flanders Fields."[50] Introduced in 1921 by the British Legion to recall Britain's war dead, the poppies were distributed each year in time for Armistice Day and continue to be manufactured today for the annual Remembrance Sunday commemorations. As Geoff Dyer reflects upon visiting the battlefields of World War I:

> So strong are these feelings that I wonder if there is not some compensatory quality in nature, some equilibrium—of which the poppy is a manifestation and a symbol—which means that where terrible violence has taken place the earth will sometimes generate an equal and opposite sense of peace.[51]

Indeed, the eternal sleep that these poppies both induce and symbolize speaks to nature's ability to restore a sublime sense of peace after the war. They grow over the battlefields naturally: No one plants them or nurtures their growth. They are ironic, organic, and in their brightness and solitude, seem to speak for themselves, requiring no additional narrative or context. That is, they serve memory in a way that is fundamentally opposed to the remembrance West carves out in "Greenhouse with Cyclamens."

The cyclamens that flourish in Nuremberg do not call up images of a natural world gradually healing itself in the war's wake, although West, as I have suggested, played with this established sense of irony. Absent their caretaker and his clientele, there would be no such images to speak of: Nuremberg would be another war-torn city without signs of the life that will, literally and metaphorically, grow over the scars of the past. Instead of using a discrete symbol, West complicates the vicissitudes of memory, and her complex web of connected experiences and images suggests that no symbol on its own will suffice to call to mind a disastrous era where grief was tended by law, where remembrance became entangled with justice—and where this justice, as West reminded her readers time and again, was meted out in the dullest of ways.

"There is not so much banality of evil," writes philosopher Avishai Margalit, "as banality of indifference."[52] In wending her way out of Nuremberg's boredom, I would suggest that West wrote precisely against this indifference. She identified in Nuremberg a fundamental problem of sensibility: a sense that the trial's dullness eroded one's capacity to feel. Those who sat in the courtroom were rendered senseless or, to put it bluntly, were bored out of their minds—and as West feared, out of their memories as well. Because this indifference steers one away from memory—why should one remember what one does not care about?—one must work around it as best one can. But "Greenhouse with Cyclamens" offered another approach: not working *around* but working *through* this boredom, harnessing it and utilizing it to shape an experience that mattered to memory.

I am not claiming this process of "working through" boredom allowed one to overcome it; in this sense, my use of the phrase parts ways with its usual sense of psychoanalytic *Durcharbeitung*. My emphasis, rather, is on boredom as a genre—and a pretext—to arrive at justice by other means. This does not imply, moreover, that what happens in the courtroom is not itself justice, but it is vastly different from the justice that occurs in the outside world. Whether we cast these differences as legal or social, explicit or subtle, monumental or minor is of little importance here. What is important, however, is that these modes of justice do not exist in opposition to each other: not legal *versus* social but legal *and* social justice as part of a larger cultural, historical imaginary in which official and unofficial judgments occur simultaneously. Although it may be tempting to identify these two forms of justice as indifferent to each other, doing so would overlook the ways in which they are, in fact, contingent: One form of justice bears on the other, exposing the other's capacities and failures.

My aim in this chapter has not been to judge the success or failure of the Nuremberg Trial but to examine what its experience consisted of and what this experience meant to the future, and the memory, of the trial. Proponents and detractors of the Tribunal have vigorously debated whether it was a fair, or merely a show, trial. And while I would not cast my lot with the Tribunal's most ardent detractors, I do believe that the trial had its limitations; I believe, nonetheless, that it was important, necessary, but by no means sufficient. My larger concern, however, has been not with the event's legal contribution but with its social and cultural legacy. I have argued for a legacy that uses the trial's boredom as a means to justice on an ordinary, social scale. Even as this more homespun justice shed light on the trial's limitations — on injustices that the court could never recognize or rectify — it also needed this official justice for these encounters to be recognized as more than just economic or political commentaries. The legal process, strictly conceived, set the stage for these encounters, but it was ultimately not the same stage on which they played out. In this process of judging outside the law, the memory of the Nuremberg Trial was itself made. And as West reminds us, what we make of it, we make ourselves.

In West's "making" of the trial, our ideas about justice are profoundly related to the way we experience an event *as it occurs*. Indeed, the tedium of Nuremberg extended some of West's private sentiments about World War II, which she, along with many others, referred to as "The Great Bore War." [53] As her most recent biographer put it, "If the bombing of London terrified and saddened people, it also bored them, as they became accustomed to the underground shelters and awaited the bombers' arrival." [54] And as West herself wrote of those times, "The air . . . beats with a faint pulse which slows down one's own . . . The mind swings loose of the present." [55] These sentiments seemed to reassert themselves in Nuremberg, where West dwelled not on the trial's innovativeness but on the fact that its experience was neither gripping nor exhilarating.

And perhaps this is part of the very reason it fell short of her expectations: It was *experienced* rather than imagined. It was this experience, one marked by West's position as a witness at the trial, that also differentiates "Greenhouse with Cyclamens" from *The Return of the Soldier*. For West, like her narrator Jenny, must imagine the experience of war in fiction and can thus accord it dramatic but always somewhat unreal proportions. Yet no such fantasy could enter into her account of Nuremberg if she was to convey the trial faithfully. To do it justice, she implied, is to record it without embellishment, with an unwavering directness.

West's account suggested, moreover, that boredom accounted for a crucial part of the Nuremberg experience and thus needed to be woven into the trial's narrative. She insisted, above all, that this tedium itself became the *purveyor* of this experience. It was thus not against but through this boredom that one could take responsibility, which for West amounted to making memory. Thus, even as she critiqued the trial for being "an unshapely event, a defective composition, stamping no clear image on the mind of the people it had been designed to impress," her work suggested that this very unshapeliness, this nonexperience, calls on us to actively remember the trial, to fashion mnemonic techniques that give it shape, amplitude, and above all enough time to wend one's way down the avenues of memory.

"Greenhouse with Cyclamens" ultimately reminds us that the most memorable moments are ironically not always those that stand out and captivate us. Indeed, the events that fall short of our expectations and threaten to fade into obsolescence also, it turns out, call on us to *work* to commit them to memory. The Nuremberg Trial may well have been dull and uninspiring. Yet the consequences of forgetting the stakes of even the dullest trial, and of leaving its details solely to the historian or the novelist, would be grim indeed. It is in this spirit that West concludes "Greenhouse with Cyclamens I" with a meditation on bravery: "Brave [were] the men," she writes, "who, in making the Nuremberg trial, tried to force a huge and sprawling historical event to become comprehensible. It is only by making such efforts that we survive" (*TP*, p. 267). For in spite of all there was to lose—foremost, perhaps, the captive audience that had gathered to see the "show" (or to read about it)—those responsible for the trial did an invaluable service to justice.

Moreover, in committing the Nuremberg Trial to memory through the ordinary, if odd, world surrounding it, Rebecca West responded to her present in a manner at once ethical and normative. If memory is ethical, it is so because of the commitments we bring to it. And if it is normative, it is because we insist on those commitments, fashioning ways to ensure that they continue to make themselves a pressing part of the present rather than a still life, a closed chapter of the past. West's account of Nuremberg suggested that this insistence amounted to more than simply articulating one's commitments. Rather than announcing the imperative to remember, she subtly and carefully crafted a landscape that steered one toward this memory. Hers is a dynamic world full of strange encounters, surprising experiences, and unlikely symbols. And yet the more one navigates its contours, the more memory is set in motion—a movement made possible because memory is given a context and a *sense* rather than a verdict.

By making the Nuremberg Trial dull, ordinary, and above all, relevant, West put forth a new sense of responsibility for history: The trial may well be legally exemplary, but it is also, in its tedium, decidedly contemporary. As such, it cannot be dismissed as an isolated experience whose uniqueness borders on irrelevance or obsolescence. It is worth remembering because, like the people in the streets of London, it may be disregarded and forgotten all too easily. But it also, as West muses in "The Dead Hand," "happen[s] to compose the age in which I live"; it is intimately bound with the concerns of everyday life and thus demands attention and regard. Seen in this light, Nuremberg's postwar sense of justice continues to be meaningful long after the court has adjourned. What emerges from West's understanding of contemporariness, ordinariness, and relevance is thus a new relationship to the past, which no longer offers a refuge from the ills of the present. Instead, the present demands our total engagement, forcing a break with the past and compelling us to face reality without the distinguished, but never quite relevant, shelter of monumental precedent.

Notes

1. Rebecca West, *The Meaning of Treason* (London: Macmillan, 1949), 70.

2. Patricia Meyer Spacks, *Boredom: The Literary History of a State of Mind* (Chicago: University of Chicago Press, 1995), 24.

3. See in this regard Marc Osiel, *Mass Atrocity, Collective Memory, and the Law* (New Brunswick, NJ: Transaction Publishers, 2000).

4. Although there were twelve trials conducted under the American occupation authorities in Nuremberg between 1946 and 1949, I refer in this chapter only to the first Nuremberg trial of the major war criminals, which is commonly referred to as the Nuremberg Trial.

5. Indeed, the decade following these trials saw the field of jurisprudence grappling with questions of law's constitution and basis. The 1958 debate between legal scholars H. L. A. Hart and Lon Fuller, which grappled with the question of whether unjust law nonetheless constituted law, illustrated the force with which the issues raised by Nuremberg—and in the case of the Hart–Fuller debate, the tension between positive and natural law—resonated in the wake of the trials. See H. L. A. Hart, "Positivism and the Separation of Law and Morals,"

Harvard Law Review 71 (1958), 593; and Lon L. Fuller, "Positivism and Fidelity to Law—A Reply to Professor Hart," *Harvard Law Review 71* (1958), 630.

6. Laurel Leff, "Jewish Victims in a Wartime Frame: A Press Portrait of the Nuremberg Trial," unpublished article, 2002. For a broader account of journalism in the United States during the Holocaust, see Leff's *Buried by The Times: The Holocaust and America's Most Important Newspaper* (Cambridge: Cambridge University Press, 2005).

7. Gloria Fromm has proposed that trials held particular appeal for West because of their limited framework, one that "provided a ready-made structure" into which she was able to integrate psychological reflections about the individuals on trial. As I will argue with respect to West's essays on Nuremberg, however, her focus shifted away from the experience of those on trial, settling instead on the experience of those who sat and watched the proceedings—thereby pushing the conventional format of trial reporting to its limits. See Gloria G. Fromm, "Rebecca West: The Fictions of Fact and the Facts of Fiction," *New Criterion 9.5* (1991), 44.

8. See, for example, Zofia P. Lesinska, *Perspectives of Four Women Writers on the Second World War: Gertrude Stein, Janet Flanner, Kay Boyle, and Rebecca West,* Studies in Literary Criticism and Theory Series 1 (New York: Peter Lang, 2002); and Brian Hall, "Rebecca West's War," *The New Yorker,* April 15, 1996, 74–83.

9. Rebecca West, undated note, Box 38, Folder 1402, *Rebecca West Papers.* General Collection of Rare Books and Manuscripts. Beinecke Rare Book and Manuscript Library, Yale University.

10. Rebecca West, *A Train of Powder* (London: Macmillan, 1955), 8.

11. Rebecca West, *Ending in Earnest: A Literary Log by Rebecca West* (Garden City, NY: Doubleday, Doran, 1931), 34–35.

12. Ibid., 38.

13. Ibid., 38–39.

14. Ibid., 38–40.

15. Rebecca West, *A Train of Powder* (London: Macmillan, 1955), 3. All subsequent references will be cited in the text as *TP,* followed by the page number.

16. Robert E. Conot, *Justice at Nuremberg* (New York: Harper & Row, 1983), xi.

17. Robert H. Jackson, "Opening Address for the United States, Nuremberg Trials," in *Philosophical Problems in the Law* (2nd ed.), David M. Adams, ed. (Belmont, CA: Wadsworth, 1996), 5.

18. Otto Fenichel, "On the Psychology of Boredom," in *The Collected Papers of Otto Fenichel: First Series* (New York: Norton, 1953), 301 (emphasis in the original).

19. "During the Nuremberg trial, the *Times* reporters fully accepted the prosecutors' version of the case, so much so that they departed from standard journalistic practices. They abandoned even the pretense of objectivity or journalistic balance, never referring to 'alleged crimes' or bothering to couch the defendants' guilt in terms of the presumption of innocence. Nor did they challenge the authenticity or reliability of the prosecution's

evidence, or suggest an alternative interpretation of witnesses' testimony." Leff, "Jewish Victims in a Wartime Frame," 8.

20. Janet Flanner, "Letters from Nuremberg," in *Janet Flanner's World: Uncollected Writings 1932–1975*, Irving Drutman, ed. (New York: Harcourt Brace Jovanovich, 1979), 109.

21. Lawrence Douglas, *The Memory of Judgment: Making Law and History in the Trials of the Holocaust* (New Haven, CT: Yale University Press, 2001), 84–85. Flanner would repeat this contrast in various ways throughout her reports, as she does in her description of the trial's ceremonial grandeur: "One of the sights outside the courthouse is the glorious daily arrival of Sir Geoffrey in a magnificent black limousine, glistening against the dusty ruins of the bolbed walls. Attired in a long, blue broadcloth coat and a bowler, he passes through the courthouse door while the Allied guards of the day—the Russians with medals or the French with berets or the Tommies with battle ribbons or the Americans in snow-white helmets—stiffly present arms. In the courtroom itself, the same physical dignity and sartorial elegance of prosecutor Sir David Maxwell Fyfe, impeccable in his Foreign Office attire, have unquestionably affected the Nazis, hypersensitive to formality and chic in the male." Flanner, "Letters from Nuremberg," 117.

22. The same lack of distinction extends West's understanding of the relationship between public and private. As Janet Montefiore has argued with regard to *Black Lamb and Grey Falcon*: "The story of Rebecca West's own journey cannot, then, be separated from the wider story of Yugoslavia, which is itself a stage for the European crisis. You cannot distinguish private from public: personal friendships express political hope and divisions, no more and no less than the mine at Trepcha, the buildings in the cities, the marketplaces. Yugoslavia is not only a border country; it is a stage where tensions of Europe are acted out, in intimacies as well as in battles." Janet Montefiore, *Men and Women Writers of the 1930s: The Dangerous Flood of History* (London: Routledge, 1996), 207.

23. Flanner, "Letters from Nuremberg," 117.

24. Douglas, *The Memory of Judgment*, 85.

25. West would continue to insist on the historian's limitations, stating in an interview in 1973, "I think it was an extraordinarily good thing that the Second World War went to the lawyers and not to the historians, because lawyers do have a certain degree of impersonality about them and they don't get so many bees in their bonnets. If you meet a lawyer he's less likely to strike you as slightly loony than historians are. I think historians are the wildest type of intellectual I have ever encountered." Anthony Curtis, "Dame Rebecca West Talks to Anthony Curtis About Social Improvements and Literary Disasters," *Listener*, February 15, 1973, 21.

26. Jackson, "Opening Address," 6.

27. As Laurel Leff reminds us, most of the coverage on the trial focused on its beginning; as the months wore on, the headline stories declined dramatically, suggesting that few people had any sense of what was going on beyond those in the courtroom. Leff, "Jewish Victims in a Wartime Frame," 11.

28. In his Introduction to Whitney Harris's history of Nuremberg, Chief Prosecutor

Robert Jackson would echo this sentiment less than a decade later: "During the almost year-long trial, it was not practicable for the daily press to present American readers with more than occasional, sketchy, and sometimes inaccurate accounts of the evidence and proceedings, nor was there in this country the wide and sustained reader-interest felt by the peoples of Europe, whose countries had been occupied. As a result, no sound and general foundation of public information about the trial was laid." Robert H. Jackson, "Introduction," Whitney R. Harris, *Tyranny on Trial: The Trial of the Major German War Criminals at the End of World War II at Nuremberg, Germany, 1945–1946* (Dallas, TX: Southern Methodist University Press, 1954; rev. ed., 1999), xxxvi.

29. Robert Jackson, "Report to the President, June 6, 1945," in *The Nuremberg War Crimes Trial 1945–46: A Documentary History*, Michael R. Marrus, ed. (Boston: Bedford Books, 1997), 40. For a probing analysis of the impact of this documentary strategy, see Chapter 1 in Douglas, *The Memory of Judgment*.

30. Victoria Glendinning, *Rebecca West: A Life* (New York: Fawcett Columbine, 1987), 194.

31. Pierre Nora, *Les lieux de mémoire* (Paris: Gallimard, 1984).

32. Quoted in Mary Carruthers, *The Book of Memory* (Cambridge: Cambridge University Press, 1990), 16.

33. A. R. Luria, *The Mind of a Mnemonist: A Little Book About a Vast Memory*, Lynn Solotaroff, trans. (Cambridge, MA: Harvard University Press, 1987), 11–12 (emphasis in the original).

34. Ibid., 31–32.

35. Virginia Woolf advances a similar approach to remembering through familiar surroundings in "A Sketch of the Past": "I see it—the past—as an avenue lying behind; a long ribbon of scenes, emotions. There at the end of the avenue still, are the garden and the nursery. Instead of remembering here a scene and there a sound, I shall fit a plug into the wall; and listen in to the past." Virginia Woolf, "A Sketch of the Past," in *Moments of Being* (San Diego, CA: Harcourt Brace, 2nd ed., 1985), 67.

36. Luria, *The Mind of a Mnemonist*, 28.

37. Jerome Bruner, Foreword to the 1987 edition, in Luria, *The Mind of a Mnemonist*, x.

38. Jerome Bruner, Foreword to the first edition, in Luria, *The Mind of a Mnemonist*, xxii.

39. Luria, *The Mind of a Mnemonist*, 116.

40. Rebecca West, *The Meaning of Treason* (London: Macmillan, 1949), 70.

41. West's statement underscores the force of form and the inadequacy of this form—both legal and artistic—in shaping the trial, and her words echo those of Jackson's opening address: "Despite the magnitude of the task, the world has demanded immediate action. This demand has had to be met, though perhaps at the cost of finished craftsmanship." Jackson, "Opening Address," 8.

42. West's approach thus draws boredom's relationship to time, as it is noted by Otto Fenichel: "The very word '*Langeweile*' indicates that in this state there are always changes

in the person's subjective experience of time. When we experience many varying stimula-tions from the outside world, the time, as we know, appears to pass quickly; but should the external world bring only monotonous stimuli, or should subjective conditions prevent their being experienced as tension-releasing, then the 'while is long.'" Fenichel, "On the Psychology of Boredom," 301.

43. Adam Phillips, *On Kissing, Tickling, and Being Bored: Psychoanalytic Essays on the Unexamined Life* (Cambridge, MA: Harvard University Press, 1993), 69.

44. Ibid., 74 (emphasis added).

45. Walter Benjamin, "The Storyteller," *Illuminations*, Harry Zohn, trans., Hannah Arendt, ed. (New York: Schocken Books, 1968), 91.

46. Her method echoes Benjamin's own commitment to withholding interpretation, which he takes as a central component in making a story at once meaningful and memo-rable: "Every morning brings us news of the globe, and yet we are poor in noteworthy stories. This is because no event any longer comes to us without already being shot through with explanation. In other words, by now almost nothing happens that benefits storytell-ing; almost everything benefits information. Actually, it is half the art of storytelling to keep a story free from explanation as one reproduces it. [. . .] The most extraordinary things, marvelous things, are related with the greatest accuracy, but the psychological connection of the events is not forced on the reader. It is left up to him to interpret things the way he understands them, and thus the narrative achieves an amplitude that information lacks." Benjamin, "The Storyteller," 89.

47. Luria, *The Mind of a Mnemonist*, 11.

48. Ibid., 12.

49. For a perceptive analysis of these images' relation to memory, see Lawrence Douglas, "The Shrunken Head of Buchenwald: Icons of Atrocity at Nuremberg," *Representations 63* (1998), 39–64; and Douglas, *The Memory of Judgment*, Chapter 1.

50. *The Great War Reader*, James Hannah, ed. (College Station: Texas A&M University Press, 2000), 368.

51. Geoff Dyer, *The Missing of the Somme* (1994; London: Phoenix Press, 2001), 130.

52. Avishai Margalit, *The Ethics of Memory* (Cambridge, MA: Harvard University Press, 2002), 34.

53. Rebecca West, letter to Mary Andrews, October 21, 1939, Box 2, Folder 32, *Rebecca West Papers*. General Collection of Rare Books and Manuscripts. Beinecke Rare Book and Manuscript Library, Yale University.

54. Carl Rollyson, *Rebecca West: A Life* (New York: Scribner's, 1996), 201.

55. Ibid., 203.

Mandating the National Memory of Catastrophe

JAMES E. YOUNG

Can the national memory of catastrophe be legally mandated? Can such memory be legislated and thereby legally proscribed? And if so, can it be legally enforced? Can an entire polis be told what to remember, how, and why? All in the name of national unity? Conversely, can other kinds of national memory be proscribed? Or is national memory a necessary fiction of the nation-state, whereby a heterogeneous polis is invited to mistake a shared memorial space, say a day or national museum, for shared memory? And finally, just how is a memorial mandate to be articulated formally in the designs of a monument, museum, or commemorative day?

Here I would like to reflect on these questions, the mandates of memory as represented in a handful of national memorial institutions and days of remembrance. But because I regard memory more as a process and a coming to terms with loss than a concretized end result, I would like to bring back into view the legislative debates by which particular memorial mandates have been negotiated. By restoring the memory of these mandates' own coming into being, I hope to reanimate them with their own dynamic genesis in real historical time. As I've done with memorials themselves, I hope this kind of restoration of the process will animate and revivify such memory, recalling that memorial laws, too, are negotiated in human time and space and that their codification by no means ends this negotiation but, in fact, invites their constant reinterpretation in new times and circumstances. Process might thus be regarded as the lifeblood of memorial mandates, with the end of process regarded as the foreclosure of memory over time.

The "memorial mandates" I would like to examine here include Israel's day of Holocaust remembrance and the U.S. Holocaust Memorial Museum. I'd like to end with a brief reflection on the attempts to mandate memory of the 9/11 attacks in New York, as found in the World Trade Center Memorial design process.

Can national memory be legally mandated? Yes, it is being legally mandated all the time. But it is always memory in the service of a national self-idealization, never in the service of recalling particular events for their own sake. By pointing this out, we don't mean to undermine any particular form of mandated memory, insofar as they are all nationally mandated and shaped. But we do want to en-sure that such memory remains contingent over time, thereby alive to new times and circumstances. By emphasizing the process of memory over its codification, the constant evolution of memory over its univocal ideal, we animate it with its own coming into being.

In fact, Maurice Halbwachs argued that it is primarily through membership in religious, national, or class groups that people are able to acquire and then recall their memories at all.[1] That is, both the reasons for memory and the forms memory takes are always socially mandated, part of a socializing system whereby fellow citizens gain common history through the vicarious memory of their fore-bears' experiences. If part of the state's aim, therefore, is to create a sense of shared values and ideals, then it will also be the state's aim to create the sense of com-mon memory as foundation for a unified polis. Public monuments, national days of commemoration, and shared calendars thus all work to create common loci around which seemingly common national identity is forged. The aims of a state's national memorial mandates, in other words, are twofold: to commemorate par-ticular events and to create a unifying sense of shared history.

At the same time, however, it has also become clear that the national need for a unified vision of the past as found in nationally mandated memorial institu-tions necessarily collides with the modern conviction that neither the past nor its meanings are ever just one thing. Because contemporary societies often perceive themselves as no longer bound together by universally shared myths or ideals, national memorial mandates extolling such universal values are often derided by late twentieth-century artists and architects as anachronistic at best and re-ductive mythifications of history at worst. How do we explain, then, the monu-ment and museum boom of the late twentieth century? The more fragmented and heterogeneous societies become, it seems, the stronger their need to unify wholly disparate experiences and memories with the common meaning appar-ently created in common spaces. But rather than presuming that a common set of ideals underpins their form, contemporary monuments often attempt to assign only singular architectonic forms to unify disparate and competing memories. In the absence of shared beliefs or common interests, memorial art in public spaces asks an otherwise fragmented populace to frame diverse pasts and experiences in common spaces. By creating common spaces for memory, monuments propagate the illusion of common memory.

Part of our contemporary culture's hunger for the monumental, I believe, is its nostalgia for the universal values and ethos by which it once knew itself as a unified culture. But this reminds us of that quality of monuments that strikes the modern sensibility as so archaic, even somewhat quaint: the imposition of a single cultural icon or symbol onto a host of disparate and competing experiences, as a way to impose common meaning and value on disparate memories—all for the good of a commonwealth. When it was done highhandedly by government regimes, and gigantic monuments were commissioned to represent gigantic self-idealizations, there was often little protest. The masses had, in fact, grown accustomed to being subjugated by governmental monumentality, dwarfed and defeated by a regime's overweening sense of itself and its importance, made to feel insignificant by an entire nation's reason for being. But in an increasingly democratic age, in which the stories of nations are told in the multiple voices of its everyday historians—that is, its individual citizens—monolithic meaning and national narratives are as difficult to pin down and mandate as they may be nostalgically longed for. The result has been a shift away from the notion of a national "collective memory" to what I have called a nation's "collected memory." Here we recognize that we never really shared each other's actual memory of past or even recent events but that, in sharing common spaces in which we collect our disparate and competing memories, we find common (perhaps even national) understanding of widely disparate experiences and our very reasons for recalling them.

Holocaust Remembrance Day

As ordered by the calendar, time offers itself as an insuperable template by which national lives are lived and past history is remembered and understood.[2] For only time, as patterned after the circular movements of Earth around the sun and the moon around Earth, can be trusted to repeat its forms perpetually. Grasped and then represented in the image of passing seasons, in the figures of planting and harvest, cycles of time have traditionally suggested themselves as less the constructions of human mind than the palpable manifestations of a natural order. As a result, both our apprehension of time and the meanings created in its charting seem as natural as the setting sun, the rising moon. By extension, when events are commemoratively linked to a day on the calendar, a day whose figure inevitably recurs, both the memory of events and the meanings engendered in memory seem as naturally true as the orbit of Earth around the sun and the moon around Earth.

As written narrative imposes the order of language onto historical events, creating in Genette's terms "narrative time," both the Gregorian solar and Jewish lunar calendars impose yet another order onto historical time, a temporal pattern by which a day's memorial significance will be understood.[3] Like historical incidents related in narrative, national commemorative days also acquire meaning according to their places on time's grid. As the placement of a monument in its city matrix generates meaning in it, the location of a commemorative day on the national calendar will create meaning in remembrance as well. When twinned with historical events, remembrance days assign specific significance to events themselves depending on where these days are located on the national calendar. The same event might carry several different kinds of significance depending on both the time of year it happened and the time of year it is remembered, which do not always coincide. Conversely, entirely disparate events acquire parallel meaning when commemorated on the same day.

In fact, perhaps nobody better recognized the potential consequences in a calendar's narrative than Israel's early state makers. The New Jews were simply going to need a New Calendar, a new time map by which to navigate the future. For the traditional calendar was as discredited in the Zionists' eyes as the self-destructive delusions it had seemed to foster in Jewish minds over the ages. It was not only a matter of rejecting 2,000 years of decayed Jewish life in the Diaspora but also of discarding the calendar that had perpetually recirculated the myths sustaining the misery of life in exile. Time would no longer be measured in the distance between the temple's destruction and the present moment. Instead, the redrawn calendar would find its genesis, its anchor, in the birth of the state itself. All else, including memorial days, would now be regarded as either culminating in Independence Day or in issuing from it.[4]

Thus declared in 1949, Israel's first and most joyous public holiday marked the Fifth of Iyar as Yom Hatzma'ut, anniversary of the day in 5708 (May 14, 1948) when independence was proclaimed and the state was established. One year later, the government dedicated the day preceding Independence Day to the memory of those who fell in the 1948 war. The choice of this date for Yom Hazikkaron initially rankled many of the bereaved families, who found such a solemn day violated by the unabashed revelry immediately following it. But the government was steadfast, its reasons for linking the state's war dead with national independence clear. On the level of pure statist ideology, no better model would be found than dying for the state: As the sole reason for living, the state would now be the only reason for dying. By yoking the deaths of its soldiers together with the birth of the state in this way, the government in effect nationalized the oldest of all Jewish paradigms: destruction and redemption. A memorial day turning at sunset into Independence Day would make explicit that the destruction of these men was redeemed in the birth of the state: Mourning was to be relieved literally by the celebration of independence.

The aim, however, was never merely to find a new worldview for the New Jew but to select which governing views to advance and which to abandon. For Ben-Gurion's statists, one of the least palatable aspects of the old calendar lay in the way its fast days had explained past disasters. Traditionally, the four minor fast days had been associated with different events from the siege and fall of ancient Jerusalem, each purportedly marking the actual anniversary of a disaster during the destruction of the Temple in 586 B.C.E. But in fact, none of them probably had any actual historical connection to the events being commemorated.[5] Nor did that matter. More important was the way that four historically independent fast days had come to be associated with the destruction of the Temple—the ur-catastrophe—and that over time, these days accumulated the commemorative weight of later catastrophes as well.

For traditional dates seem to have attracted commemoration of other events and then organized them around a teleological locus, creating a single meaning in all events. Remembrance days of multiple disasters sprang not from the coincidental occurrence of events on the same day but from the assuredly noncoincidental single meaning assigned by the tradition to all disasters, no matter how disparate. As a result, not only were entirely unrelated disasters reported to have occurred on exactly the same day, centuries apart (e.g., the destructions of the First and Second Temples),[6] but all disasters were assigned the same meaning—Mipnei Hataeinu (because of our sins). According to the Rambam, the aim of the four fasts is to "stir the hearts, to open roads to repentance, and to remind us of our own evil deeds, and of our fathers' deeds which were like ours, as a consequence of which these tragic afflictions came upon them and upon us."[7]

For the founders of modern Israel, such meaning created in the Shoah by the traditional calendar was repulsively unacceptable. That is, the religious mandate for memory was perceived by members of the Knesset to be in direct conflict with the national mandate for Holocaust memory. The first movement toward a national Holocaust remembrance day, therefore, came in the movement away from traditional commemorative dates marking former disasters. This is also where the needs of the state and those of the rabbinate would come into direct conflict. On the one side, there was a pressing need among the religious community in and out of Israel for a rabbinical ruling to set a day on which Jahrzeit candles could be lighted and kaddish recited for those whose actual dates of death during the Shoah were unknown. For the rabbinate, which date to choose was relatively simple: After all, they were already in possession of at least four ready-made days of mourning. So in 1948, they adopted the Tenth of Teveth as Yom Kaddish Klali (Day of Communal Kaddish), for little better reason, according to some, than to reinvigorate an otherwise dormant fast day.[8]

On the other hand, the obvious problem for the state and predominantly non-religious population was that according to the tradition, this day would not merely link the Shoah to the fall of Jerusalem. But in so doing, it would suggest as well the theological reasons for this fall—Mipnei Hataeinu—all as a justification for current exile. None of this could be tolerated by a state dedicated to the rebuilding of Jerusalem and the Jews' mass return from exile. If there was a "meaning" for the Holocaust in the national view, it was the necessity for a Jewish state to protect Jews from just this sort of destruction, not a divine punishment for supposed sins committed. But even here, the archaic religious paradigm is occasionally, if reflexively, reinvoked also as a Zionist explanation of events. As Liebman and Don-Yehiya have noted, the lead story in Davar on Yom Hashoah in 1960 concluded that the "function of Holocaust Day is to remind the Jewish people of its own sin in not unequivocally having chosen Zion."[9]

In fact, once recast in the image of the statists themselves, this day would not commemorate mere destruction at all. In keeping with their vision of a new, fighting Jew and their rejection of the old, passive Jew as victim, the founders would prevent this day from entering the commemorative cycle of destruction altogether. That is, not only would national memory have to be mandated separately from religious memory, but the national meaning of this memory would also have to be mandated, each as a function of the other. Six years after the liberation of the concentration camps, three years after the state of Israel was founded, the Israeli parliament thus moved to adopt a national Holocaust and Ghetto Uprising Day. As proposed by member of Knesset Mordechai Nurock on April 12, 1951, "The first knesset declares and determines that the 27th day of the month of Nissan every year shall be Holocaust and Ghetto Uprising Day [Yom Hashoah Umered

Hageta'ot]—an eternal day of remembrance for the House of Israel." [10] It remains significant that Nurock, an orthodox Jew, would agree to sponsor this bill. In an impassioned floor speech bristling with allusions to the Book of Lamentations, he was able to validate such a bill at both the political and religious levels. Not Nebuchadnezzar, nor Titus, nor the crusades or the pogroms could compare, he proclaimed. This was a third Hurban, greater than all the rest and so demanded its own day (p. 1656).

"We need to choose a date," he continued, "that coincides with most of the slaughter of European Jewry and with the ghetto uprisings that took place in the month of Nissan" (p. 1656). In the next sentence, Nurock added that the Knesset commission had chosen this day because "it was during the Sfirah, when the crusaders, ancestors [*avot avotehem*] of the Nazis destroyed so many 'holy' [i.e., Jewish] communities" (p. 1656). Since the Sfirah period (counting the omer) was already a traditional time of semimourning, during which marriages, haircuts, and music were forbidden to the religious, Nurock deemed it all the more appropriate. In addition, the only other secular date put forth until then, that of the Warsaw Ghetto Uprising (April 19, 1943, the 15th of Nissan), would not have been allowed by the rabbinate in its overlap of Passover. In fact, the ultra-Orthodox delegation to the Knesset requested that the entire month of Nissan be protected from the violation of an official day of mourning—a demand vociferously rejected by former ghetto fighters who wanted to place the day as close as possible to the anniversary of their uprising. In effect, this left only a few choice days during a period bordered on the one side by the 15th of Nissan (the first day of Pesach and Uprising) and on the other by the Fifth of Iyar, Israel's Independence Day. Forbidden to set a day of mourning during Hol Hamo'ed (the week of Passover) and not wanting to crowd Yom Hazikkaron and Yom Hatzma'ut, the committee was left with only twelve days in which to place Yom Hashoah Umered Hageta'ot.

In the end, by placing the day on the 27th of Nissan (five days after the end of Hol Hamo'ed, seven days before Yom Hazikkaron), the committee dramatically emplotted the entire story of Israel's national rebirth, drawing on a potent combination of religious and national mythologies. Pulled from both the middle of the six-week ghetto uprising and the seven-week Sfirah, this day retained links to both heroism and to mourning. In coming only five days after the end of Passover, Yom Hashoah Umered Hageta'ot extended the festival of freedom and then bridged it with the national Day of Independence. Beginning on Passover (also the day of the Warsaw Ghetto Uprising), continuing through Yom Hashoah, and ending in

Yom Hatzma'ut, this period could be seen as commencing with God's deliverance of the Jews and concluding with the Jews' deliverance of themselves in Israel. In this sequence, biblical and modern returns to the land of Israel are recalled; God's deliverance of the Jews from the desert of exile is doubled by the Jews' attempted deliverance of themselves in Warsaw; the heroes and martyrs of the Shoah are remembered side by side (and implicitly equated with) the fighters who fell in Israel's modern war of liberation; and all lead inexorably to the birth of the state.[11]

Unfortunately for this resolution, it was passed at the height of statist influence in Israel, and so, effusive parliamentary sentiments notwithstanding, was widely ignored. On the one hand, it could be argued that memory of the Shoah was not neglected so much as merely subsumed in the greater task of state building during the early 1950s. But on the other hand, this was also a time in Israel when bare mention of the Shoah, or the fact that one had survived it, might have been met with surly contempt. It was a time when survivors were still being shamed into silence by those claiming the foresight to have left Europe before the onslaught. In the early statists' view, the Shoah was redeemable—hence, memorable—by little more than instances of heroism and the Jewish courage it evoked in some of its victims, the hopelessness of Jewish life in exile, and the proven need for a state to defend Jews everywhere.[12]

Although this resolution was passed by the Knesset in 1951, it generated little public notice until 1953, when as part of its mandate, Yad Vashem Memorial Authority was assigned control over how the day was to be observed. But even then and for the next several years, it seemed the day of remembrance itself was forgotten by all but survivors and partisans. In response, Nurock and others decided in 1959 that only a law, not just another parliamentary resolution, could guarantee—that is, enforce—public observance. Some of the questions arising during the Knesset floor debate on the law's wording might now be reintroduced to the day of remembrance. First, the name: One member of Knesset found the notion of ghetto uprising too specific. After all, she asked, wasn't there also heroism in Kiddush Hashem—that is, in martyrdom itself?[13] She then proposed that since the main principle at hand was what the uprising stood for—that is, heroism—this principle should be included in the name as well: Yom Hashoah, Hagvurah, Uhamered (p. 1387). Others wondered whether we ought to remember only the killing, only the uprising, or only heroism. Why a day to remember all these? And what about the killers? How do we remember them? Or maybe this day should mark the inquisition as well. Or Chmielnicky. Or the Ukrainian pogroms.

Still others argued unsuccessfully for incorporating the day into Tesha B'av (the Ninth of Av), the most widely observed fast day recalling the destruction of the Temple. As Saul Friedlander and others have noted, years later Menachem Begin asked that Yom Hashoah Vehagvurah be divided between two already existing days. Yom Hashoah (Holocaust Day) would be observed on Tesha B'av, and Yom Hagvurah (Heroism Day) on Memorial Day for Israel's fallen soldiers.[14] Begin's proposal was primarily a reflection of Rabbi Joseph B. Soloveitchik's own proposition that the religious community would give up the Tenth of Teveth if the state gave up the 27th of Nissan—and all would commemorate the Shoah on Tesha B'av. But such a compromise would have been no compromise at all, of course, because the meaning engendered on this fast day would have been basically the same as that created on the Tenth of Teveth.[15]

So the 27th of Nissan prevailed and was eventually generalized slightly in name and assigned concrete observances. As finally passed by the Knesset on April 7, 1959, the law for this Day of Remembrance of Holocaust and Heroism reads:

1. The 27th of Nissan is the Day of Remembrance of the Disaster and Heroism, dedicated every year to remembrance of the catastrophe of the Jewish people caused by Nazis and their aides, and of the acts of Jewish heroism and resistance in that period. Should the 27th of Nissan fall on a Friday, the Day of Remembrance shall be marked on the 26th of Nissan of that year.

2. On the Day of Remembrance there shall be observed Two Minutes Silence throughout the State of Israel, during which all traffic on the roads shall cease. Memorial services and meetings shall be held in Army camps and in educational institutions; flags on public buildings shall be flown at half mast; radio programmes shall express the special character of the day, and the programmes in places of amusement shall be in keeping with the spirit of the day.

3. The Minister authorized by the government shall draft, in consultation with the Yad Vashem Remembrance Authority, the necessary instructions for the observance of the Day of Remembrance as set forth in this Law.[16]

In 1961, a further amendment required that all places of entertainment be closed on the Eve of Yom Hashoah VeHagvurah. Yad Vashem also suggested that the siren for Yom Hazikkaron be sounded to enforce the two minutes of silence on Yom Hashoah, yet another move linking the martyrs of the Shoah to the heroes who fell for the state.

Of course, to this day, enforcing such a law remains very difficult. In the ultra-Orthodox neighborhoods, the law is ignored altogether, precisely because it is a state law and not a religious law. In fact, a mandate to observe such a law of re-membrance may not necessarily include a mandate actually to remember (how is one forced to remember anything?), so much as it will be a mandate to remember something in a particular way. That is, observing the outward protocol of remem-brance may be very different from actually remembering events as prescribed. When legislators passed this remembrance law, they included no provisions (e.g., fines or imprisonment) for individual disregard of the law. They hoped that the social and political unity generated in such a law would suggest a generally agreed upon code of behavior, even decorum, during the two-minute siren. Whether this decorum is observed or not depends on the social composition of any given neigh-borhood. Ultra-Orthodox and Palestinian neighborhoods where the law is not embraced as a constitutive part of national identity therefore see less observance than in secular and Zionist neighborhoods.

How then is the remembrance day publicly performed? What do people re-member in its ceremonies and moments of silence? To what extent do the forms of observance shape remembrance itself? Rather than cutting across all locations and communities, answers to these questions depend very much on the specific site. Outside Israel, Yom Hashoah increasingly assumes the trappings of a "holy day" and so is often observed in and around the synagogue. When conducted at civic centers or at public memorial sites, "services" often are not led by a rabbi or member of the religious community.

In America, where the main organizing ideology is pluralism, ecumenical ceremonies bring together clergy from diverse faiths and ethnic groups, Jewish survivors and Christian liberators. Each commemoration reflects the ethos and tradition, the piety or politics of a given community. In fact, the first national Days of Holocaust Remembrance in America proposed by Senator John Dan-forth (April 18–29, 1979) were to commemorate the thirty-fourth anniversary of Dachau's liberation by American troops—an explicit reflection of America's Holocaust experience. Only later were these days moved to coincide more closely with the 27th of Nissan, the nearest Sunday on the Gregorian calendar.

In fact, how Yom Hashoah will remember the Holocaust to these young soldiers is made explicit in a bulletin specially prepared for army commanders on Yom Hashoah, *Informational Guidelines to the Commander*:

> 1. The Zionist solution establishing the State of Israel was intended to provide an answer to the problem of the existence of the Jewish people, in view of the fact that all

other solutions had failed. *The Holocaust proved, in all its horror, that in the twentieth century, the survival of Jews is not assured as long as they are not masters of their fate and as long as they do not have the power to defend their survival.* (emphasis added)

2. A strong State of Israel means a state possessed of military, diplomatic, social, and economic strength, and moral character which can respond properly to every threat from outside and provide assistance to every persecuted Jew wherever he is. The consciousness of the Holocaust is one of the central forces which stand behind our constant striving to reach this strength and behind the solidarity and deep tie with Diaspora Jewry.[17]

This conception further was elaborated in Chief of Staff Mordechai Gur's address at the base of Yad Vashem's Wall of Remembrance in 1976, in which he made clear the current generation's debt to the Holocaust: "If you wish to know the source from which the Israeli army draws its power and strength, go to the holy martyrs of the Holocaust and the heroes of the revolt . . . The Holocaust [. . .] is the root and legitimation of our enterprise."[18] The vocabulary of both the guidelines and the chief of staff is aimed at young soldiers, whose identity with the victims depends almost entirely on the martyrs now presented as heroes. The martyrs are not forgotten here but are recollected heroically as the first to fall in defense of the state itself.[19]

At this point, however, we might distinguish between unified forms of commemoration and the unification of memory itself, between unified meanings and unified responses to memory. For despite unified forms of commemoration, memory in these shared moments is not necessarily shared but, in fact, varies distinctly from person to person. This is not a day of shared memory but rather a shared time of disparate remembrance. Taken together, these discrete memories constitute the *collected*, not collective, memory evoked on Yom Hashoah. In this light, we might see Yom Hashoah not as a day of national memory as much as a nationalization of many competing memories.

In its conception, Yom Hashoah was intended as neither a fast nor a holy day. It was pulled out of the religious continuum precisely to be observed as a national day of remembrance. As such, this day would mirror one of the nation's own functions: to bind into one polity a diverse people. Nations traditionally accomplish this unification in a number of ways, including the propagation of common laws, ideals, and language. As it turns out, generating a national memory is yet another way to unify a nation. For the very act of commemoration provides a common experience for a population otherwise divided by innumerably disparate lives. This is not a unity of Holocaust experience, however, or even the unification of memory itself. Rather, it is only the unity of shared ceremony, which creates the sense of a shared past.

Of all memorial centers in Israel, only Yad Vashem Martyrs' and Heroes' Remembrance Authority bears the explicit imprimatur of the state. Conceived in the throes of the state's birth and building, Yad Vashem would be regarded from the outset as an integral part of Israel's civic infrastructure. As one of the state's foundational cornerstones, Yad Vashem would both share and buttress the state's ideals and self-definition. As such, Yad Vashem has sought not only to mandate memory but also to enact the state's double-sided memory of the Holocaust here.

Yad Vashem functions as a national shrine to both Israeli pride in heroism and shame in victimization, a place where Holocaust history is remembered as culminating in the very time and space now occupied by the memorial complex itself. As if trying to keep pace with the state's own growth, Yad Vashem has continued to expand its reservoir of images, sculptures, and exhibitions: As the state and its official memory of the Holocaust evolve, so too do the shapes and mandates of memory at Yad Vashem. In its role as the national "memorial authority," Yad Vashem is the final arbiter of both Holocaust memory in Israel and the very reasons for memory.

According to Shmuel Spector, the genesis for a Holocaust memorial institution in what was still Palestine came just as the first reports of the unfolding catastrophe reached the Yishuv.[20] At a board meeting of the Jewish National Fund in September 1942, Mordechai Shenhavi from Kibbutz Mishmar Ha'emek proposed to commemorate both what he called the "Shoah of the Diaspora" and the participation of Jewish fighters in the Allied armies. Shenhavi suggested that the site be called "Yad Vashem" (literally, "a monument and a name"; figuratively, "a monument and a memorial"), after a quotation from Isaiah (56:5), in which God declares how he will remember those who keep his covenant: "I will give them, in my house and within my walls, a monument and a name, better than sons and daughters. I will give them an everlasting name that shall never be effaced."

With the Yishuv itself under Nazi threat, however, the proposal lay dormant until the very last days of the war, May 1945, when the scope of destruction became clear. Resubmitting his proposal, now named "Yad Vashem Foundation in Memory of Europe's Lost Jews: An Outline of a Plan for the Commemoration of the Diaspora," Shenhavi had from the start conceived of a multidimensional site by which the Yishuv would remember both the Holocaust and former Jewish life in exile, each in the figure of the other. A month later, the Jewish National Council recommended the establishment of such an institution in Jerusalem, which would include an eternal flame for the victims; a list of their names; a memorial for the lost Jewish communities; a monument for the fighters of the ghettos; a memorial tower to honor Jewish fighters in Allied armies; a permanent exhibit on the concentration and extermination camps; and a tribute to the Gentile rescuers of Jews. Two years later, on June 1, 1947, Yad Vashem convened its first plenary session, putting on public display a plan entitled "Yad Vashem for the Diaspora."

Once again, however, implementation of the memorial was interrupted by events as dire as those that would be commemorated, in this case the 1948 War of Independence. It was not until 1950 that Shenhavi resumed his lobbying on behalf of a national memorial authority, when he got everybody's attention by proposing an audacious and startling law. Shenhavi hoped not merely to register all the victims of the Shoah, which would be to fulfill the traditional Jewish mandate to remember the dead by naming them; but now that the Jews had a state of their own, he proposed granting honorary, posthumous citizenship to all the martyrs as well. For the next two years, the memorial project was put on hold while Israel's best legal minds wrestled with the concept of commemorative citizenship. Unable to extract a ruling one way or another, the government decided to move ahead nevertheless and defer the issue of citizenship until the memorial authority itself was established.

In 1952, the minister of education and eminent Israeli historian Ben-zion Dinur submitted to the parliament a bill for the establishment of Yad Vashem. As always in the Knesset, debate was long and tangled, though the bill itself enjoyed an almost unheard-of consensus among Israeli lawmakers.[21] On May 18, 1953, spurred on by the imminent unveiling of a "memorial to the unknown Jewish martyr in Paris," the Knesset unanimously passed what was officially called the "Law of Remembrance of Shoah and Heroism—Yad Vashem" (Hok hazikaron hashoah vehagvurah—yad vashem), after which the entire assembly rose for a minute's silence in memory of the victims. On August 19, three months later and one day after the Paris memorial's unveiling, the law passed its final reading and became the first remembrance law of the land.

In its immediate temporal context, in fact, the link between the Holocaust and the establishment of the state was palpable for legislators in ways lost to and occasionally denied by subsequent generations. This was partly the result of national independence following liberation of the camps by three years, as well as a sense that Israel's War of Independence was fought as an extension of the Jews' struggle for survival in Europe. In the words of Nachum Goldman, former president of the World Jewish Congress, "If the State came into being, it was not only by virtue of the blood spilt by those who fell in the battles for its existence, which is the highest price, but also, indirectly, because of the millions murdered in the Holocaust."[22] In this view, the blood spilled for Israel's independence is seen to mingle with that spilled in Europe's slaughterhouses. If the state came about by virtue of the blood spilled in both places, it is little wonder that the murdered Jews of the Holocaust would be conferred posthumous Israeli citizenship, for in this scenario, they too have given their lives for Israel. Conversely, once martyrs of the Holocaust are united with those who fought and died for the state, the War of Independence itself might be said to have begun not in 1947 but in 1939.

Even more significant to many of Israel's leaders at the time, however, was the overt political cause and effect between the Holocaust and the UN vote for Israeli statehood. For even in the practical side of its birth, the state of Israel was tied closely to other nations' perceptions and recent memory of the Holocaust. To persuade the UN commission appointed to study the partition of Palestine into Jewish and Arab states, Goldman reported, Abba Eban and David Horowitz spent much of their time recalling to the delegates the story of the Holocaust. In their retellings, they were able to establish a persuasive link between the Holocaust and statelessness, between rehabilitation and national rebirth. As a consequence, the UN commission visited D.P. camps in Germany on fact-finding missions to determine the depth of Zionist commitment among survivors—and came away stunned by what they had seen and heard. According to several accounts, every last one of the survivors they interviewed had insisted that after Treblinka, Bergen Belsen, and Auschwitz, their future existed only in Palestine (p. 149). Even the Soviet delegate Vishinsky is reported to have risen at a meeting of his Eastern bloc comrades and declared flatly that the main factor in their deliberations was the Holocaust, for which they owed the Jews "this measure of rehabilitation."

All of this was fresh in the minds of legislators as they sought to embody the link between the Holocaust and statehood in what would become Israel's preeminent national shrine. As defined by the Martyrs' and Heroes' Remembrance Law of 1953, the memorial at Yad Vashem thus commemorates a reflexively Israeli understanding of the Holocaust, including "the six million members of the Jewish people who died a martyr's death at the hands of the Nazis and their collaborators"; "the communities, synagogues, movements . . . and cultural institutions destroyed"; "the fortitude of Jews who gave their lives for their people"; "the heroism of Jewish servicemen, and of underground fighters"; "the heroic stand of the besieged ghetto population and the fighters who rose and kindled the flame of revolt to save the honor of their people"; "the sublime, persistent struggle of the masses of the House of Israel, on the threshold of destruction, for their human dignity and Jewish culture"; "the unceasing efforts of the besieged to reach Eretz Israel in spite of all obstacles, and the devotion and heroism of their brothers who went forth to rescue and liberate the survivors"; and "the high-minded Gentiles who risked their lives to save Jews." [23]

Unlike memorials that attempt to remove their national origins and interests from view, Yad Vashem's mission as simultaneous custodian and creator of national memory was explicitly mandated in its law. Among additional tasks of Yad Vashem, as defined in Article 2, are "to collect, examine, and publish testimony of the Holocaust and the heroism it called forth and to bring home its lesson to the people"; "to promote a custom of joint remembrance of the heroes and victims"; and "to confer upon the members of the Jewish people who perished in the days of the Holocaust and the resistance the commemorative citizenship of the state of Israel, as a token of their having been gathered to their people" (p. 5). Shenhavi would have his wish. Among the early tasks of Yad Vashem was the Registration Project, launched April 19, 1955 (Uprising Day), to record the names of every Jewish victim of the Germans. As defined in the *Yad Vashem Bulletin*, the purpose of the Daf-Ed (Memorial Page) project is "to perpetuate the memory of the millions of martyrs, whose graves are unknown and unmarked, by registering their names and other particulars in 'Memorial Pages' and awarding them 'memorial citizenship' of the State of Israel." [24] In bestowing posthumous citizenship on the victims, the state effectively created an invisible but ever-present shadow population of martyrs. As the state's newest citizens, these martyrs are understood in this context as having been murdered not only because they were Jews but because they were Israelis as well.

The function of the memorial mandate here is precisely what it has always been for the Jewish nation; that is, in addition to bringing home the "national lessons" of the Holocaust, this law would work to bind present and past generations, to unify a world outlook, to create a vicariously shared national experience. These are the implied functions of every national memorial, of course, merely made visible in Israel's legislation of such memory.

From the beginning, almost every year has witnessed another unveiling at Yad Vashem of a new memorial sculpture or gardens placed around the grounds, including reproductions of memorial sculptures from the Warsaw Ghetto and Dachau. A monument and plaza commemorating Jewish soldiers in the Allied forces were added in 1985, and a children's memorial was added in 1988. A memorial sculpture commemorating four martyred women, heroines of the Auschwitz Sonderkommando uprising, was dedicated in 1991. A huge project, "The Valley to the Destroyed Communities," was completed in 1992. In all cases, however, Yad Vashem concerned itself with only the destruction of Jews during the war—not with other groups murdered en masse by the Nazis. Even so, the construction of memory at Yad Vashem has spanned the entire history of the state itself, paralleling the state's self-construction. For this reason, it seems clear that the building of memorials and new spaces will never be officially completed and that, as the state grows, so too will its memorial undergirding.

Indeed, as the state has begun to recognize the fact of its plural and multiethnic society, and as it recognizes its own debt to globalization, its perception of the Holocaust has also begun to evolve to include other than Jewish victims of the Nazis. With a new generation's mandate in mind, Yad Vashem has thus completely revamped its historical exhibition to reflect a new generation's reasons for remembering such history in the first place. The most significant of the many changes in Yad Vashem's new historical museum, therefore, is a narrative that includes not just Jewish victims of the Nazis but also the Sinti and Roma. Jehovah's Witnesses, political prisoners, homosexuals, and even Polish clergy and German victims of the Nazis' early T-4 (euthanasia) program for the mass murder of people with disabilities and handicaps. In a land of immigrants—including Christian Russian spouses of Jewish immigrants from the former Soviet Union and Ethiopian Jews—and in a time when young people are increasingly looking outward at other groups of contemporary victims in the world around them, Yad Vashem also now sees the need to tell the stories of victims other than Jews. With a newfound grasp of itself as a plural, immigrant nation, Israel's national institutions have begun to negate the traditional Zionist negation of the Diaspora.

Indeed, at this very moment, a furious debate is underway in Israel between those who hold that Israel's past must remain inviolate as a unifying experience that justifies Israel's very reason for being and those who believe Israel is now strong enough to reevaluate its sacred national myths of origin toward a more complete and balanced understanding of the Palestine–Israel conflict. Although two of Israel's "new historians"—Benny Morris and Zev Sternhel—have concentrated primarily on reexamining Israel's wars and the Palestinian conflict overall, two other "new historians"—Moshe Zuckerman and Idit Zertal—have looked closely at the ways traditional national memory of the Holocaust has shaped Israel's understanding of its wars and the Palestinian conflict.

Rather than paraphrase the new historians here, I would like to cite the actual words of a well-respected Israeli historian of science and former director of the Van Leer Institute in Jerusalem, Yehudah Elkana, from an article he wrote for *Ha'aretz*, which appeared March 2, 1988—and which I believe explicitly defined the terms of the ensuing debate. Entitling his article "In Praise of Forgetfulness," Elkana urged the nation's leaders to let go of the memory of the Holocaust and "to position themselves in favor of life, to dedicate themselves to the building of the future and to cease dealing morning and evening with the symbols, ceremonies, and lessons of the Holocaust." At the time he wrote this piece, the country was embroiled in the first Intifada and was awaiting a verdict in the trial of John Demjaniuk, supposedly Ivan the Terrible of Treblinka. In fact, it was precisely this national juxtaposition of Holocaust memory and the first Palestinian uprising that moved Elkana to write. For even though Elkana himself was a survivor of Auschwitz, he had come to feel that commemorating the Holocaust in Israel was no longer about mourning past Jewish victims or even about reminding all about how and why the state itself had come into being. But he now felt that Holocaust memory had turned into a justification for extreme violence against any and all of Israel's enemies, both real and imagined. In his words,

> Lately, I am more than ever convinced that it is not a personal frustration as a political social motive driving Israeli society in its relationship with the Palestinians, but a deep existential anxiety which is nurtured by a certain interpretation of the Holocaust's lessons and by the willingness to believe that the entire world is against us and that we are the eternal victim. This age-old belief, which is shared by so many is, to my view, Hitler's tragic and paradoxical victory.

Elkana went on to describe the obsessive memory of past suffering as a threat to democracy itself in Israel. He concluded, "Maybe the world should remember the Holocaust, though I am not even convinced of this. Anyhow, it is not for us to worry about it. Every people, the Germans included, will decide about their own ways of remembering. We, on the other hand, must forget."

With these words, the opening shot was fired, and a postwar generation questioning Israel's national mandate for Holocaust memory now found its voice. In fact, the new historians have largely succeeded both in recasting Israel's textbooks and in so doing recasting Israel's own perception of its enemies, both current and past.

One recent book in particular, Idit Zertal's *From Catastrophe to Power*, goes back to analyze step by step what she calls Israel's "mobilization of Holocaust memory"—from the birth of the state onward.[25] Although the line between the deliberate instrumentalization of the Shoah and the more reflexive, even automatic assimilation of such memory into the present moment may not be as stark and well defined as Zertal suggests, she does unpack this unholy mix of instrumentalization and the reflexive use of Holocaust memory in Israel's fraught political culture. But at her harshest, she may attribute a malevolence in intent to Israel's founders that neglects the more highly contingent circumstances of unfolding events at the time. Did Ben-Gurion, for example, know that his language of Holocaust redemption in the state of Israel would be extended to sanctifying occupied territories later? Or is it possible that the politicians' worldviews were actually as conditioned by the teleological equation—me'shoah le'geulah—as their Holocaust memorial rhetoric was?

When Ben-Gurion dismissed two millennia of Jewish history outside Israel as barely worth recalling, was it out of callous indifference to the victims of the Shoah or the survivors? Or was he merely reciting, mantra-like, the Zionist negation of the Diaspora? Among other insights, Zertal shows just how blinkered even the greatest intellects of Zionism were by their very ideology: a critique of the Diaspora in which life in exile was by definition neither worth living nor remembering. In fact, as Zertal suggests, it may not be memory per se that is the problem but only certain *kinds* of memory, which are indeed a problem. Any mandating of Holocaust memory that leads to rationalizing the occupation of the territories is for Zertal and her cohort clearly a mandate that oppresses life.

Some wonder whether such doubt ever makes a state stronger. After forty years or so, with the advent of Israel's new historians, I happen to believe that Israel's new historians are not only proof of a strong and confident national culture, but in their disentangling of Israel's gilded myths of origins from the more complex facts on the ground, I believe they make both the state and its culture much stronger still. Obversely, I find that national myths in the service of national founding become dangerous only if they are believed literally by succeeding generations—who must, after all, not only act on real events in real time but must also define and create their *own* national, even mythological reasons for being. This is why the early work of new historians like Benny Morris and Idit Zertal, among others, is so important. For only by clearing away the brush of obfuscating national myths and recalling the events these myths have obscured, they now make way for a new, ever evolving rationale for national existence.

Indeed, with Israel's new historians as our model, we might see ourselves not in a period of post-Zionism as much as in a time of "new Zionism." This renewed Zionism, if you will, would be the continued belief in the right of Jews to a national home. But it would be rooted no longer in the state's founding myths as much as in the conviction not only that—m'dor l'dor—every new generation has to find its own national reason for being but that memory of our national origins recognizes its debt to both history and to the present moment. In this way, I find that Israel's national memory is being expanded to include not just the history of Jews but also the histories of those around us—such as the Palestinians—whose lives have been so indelibly shaped *by* Jewish history. To the extent that Israel's, or any nation's, memorial mandate responds and adapts itself to new historical and political realities, it is sustained and even vivified by time and change and not made obsolete by it.

American National Memory of the Holocaust

The first public Holocaust commemoration in America took place not after the war at all but at the very height of the killing, December 2, 1942, as a mass protest. On this day, according to the Jewish Telegraphic Agency, some 500,000 Jewish workers in New York City stopped work for ten minutes, both to mourn those already killed and to protest the ongoing massacre. In a gesture of sympathy, several radio stations observed a two-minute silence before broadcasting memorial services at 4:30 that afternoon.[26] Similar commemorations followed the next spring, culminating in several mass public memorial ceremonies, including a pageant held at Madison Square Garden in March 1943, called "We Will Never Die," and dedicated to the 2 million Jews who perished at the hands of the Germans that year.[27] Other public memorials included mass rallies called by the Jewish Labor Committee to mourn the destruction of the Warsaw ghetto. The largest single Holocaust memorial event during the war took place on April 19, 1944, the first anniversary of the Warsaw Ghetto Uprising. On the steps of New York City Hall, over 30,000 Jews gathered to hear Mayor Fiorello LaGuardia and prominent Jewish leaders honor the memory of fighters and martyrs who died in the uprising one year before.

At another massive public ceremony in October 1947, the next mayor of New York, William O'Dwyer, dedicated a site in Riverside Park between 83rd and 84th Streets and marked it with a plaque reading: "This is the site for the American memorial to the Heroes of the Warsaw Ghetto Battle, April–May 1943 and to the six million Jews of Europe martyred in the cause of human liberty." The plaque, with its characteristically American emphasis on being "martyred in the cause of human liberty," remains to this day, but the memorial was never built, its mandate never ratified by the city. As is often the case, the subsequent story surrounding the unbuilt memorial can be more instructive than the finished memorial could ever have been.

In 1964, when a group of Jewish American survivors of the Warsaw Ghetto Uprising submitted a design for a Holocaust memorial at this site to New York City's Arts Commission, they were turned down for three reasons. First, according to Arts Commissioner Eleanor Platt, the proposed design by Nathan Rapoport was simply too big and not aesthetically tasteful. It would, in her words, set "a regrettable precedent." Second, such a monument might inspire other "special groups" to be similarly represented on public land—another regrettable precedent. And finally, according to Parks Commissioner Newbold Morris, the city had to ensure that "monuments in the parks . . . be limited to events of American history."[28] That is, he suggested, the Holocaust was not an American experience.

For the Jewish survivors of the Holocaust who had immigrated to America after World War II, and who regarded themselves as typical "new Americans," such an answer challenged their very conception of what it meant to be American in the first place. For the first time in their minds, a distinction had been drawn between "events of American history" and those of "Americans' history." Did American history begin and end within the nation's geographical borders? Or did it, as most of the survivors believed, begin in the experiences abroad that drove these immigrants to America's shores? With the 1993 dedication of the U.S. Holocaust Memorial Museum in Washington, D.C., it could be said that America has recognized the survivors' experiences as part of a national experience—and has in this way made the Holocaust part of American history. American memory might now also be said to include the memories of all Americans.

At the same time, such a museum necessarily raises other difficult questions, among them: What role does the Holocaust play in American thought and culture, in American religious and political life, in relations between Jewish Americans and other ethnic groups? To what extent will it necessarily be universalized in a society defined by pluralist and egalitarian ideals? To what extent has it become a defining preoccupation for Jewish Americans, a locus of memory and identity? Can there even be a single national mandate for Holocaust memory in such a heterogeneous society? The answers to these questions, and hence the memorial mandate itself, are complicated and ever changing.

In America, the motives for memory of the Holocaust are as mixed as the population at large, the reasons variously lofty and cynical, practical and aesthetic. Some communities build memorials to remember lost brethren, and others to remember themselves. Some build memorials as community centers, and others as tourist attractions. Some survivors remember strictly according to religious tradition, whereas others recall the political roots of their resistance. Veterans' organizations sponsor memorials to recall their role as camp liberators. Congressmen support local monuments to secure votes among their Jewish constituency. Even the U.S. Holocaust Memorial Museum in Washington, D.C., was proposed by then President Jimmy Carter to placate Jewish supporters angered by his sale of F-15 fighter planes to Saudi Arabia. All such memorial decisions are made in political time and are contingent on political realities.[29]

Of all the Holocaust memorials in America, none can begin to match in scope or ambition the national memorial and museum complex that opened in April 1993 in the heart of the nation's capital. Situated adjacent to the National Mall and within view of the Washington Monument to the right and the Jefferson Memorial across the Tidal Basin to the left, the U.S. Holocaust Memorial Museum is a neighbor to the National Museum of American History and the Smithsonian Institute. It has enshrined, by dint of its placement, not just the history of the Holocaust but American democratic and egalitarian ideals as they counterpoint the Holocaust. That is, by remembering the crimes of another people in another land, it encourages Americans to recall their nation's own idealized reason for being.

"What is the role of [this] museum in a country, such as the United States, far from the site of the Holocaust?" Charles Maier has asked. "Is it to rally the people who suffered or to instruct non-Jews? Is it supposed to serve as a reminder that 'it can happen here?' Or is it a statement that some special consideration is deserved? Under what circumstances can a private sorrow serve simultaneously as a public grief?"[30] Before such a museum could be built on the mall in Washington, D.C., explicitly American reasons would have to be found for it.

The official American justification for a national memorial in the nation's capital was provided by President Jimmy Carter in his address to the first Days of Remembrance commemoration in the Capitol Rotunda on April 24, 1979:

> Although the Holocaust took place in Europe, the event is of fundamental significance to Americans for three reasons. First, it was American troops who liberated many of the death camps, and who helped to expose the horrible truth of what had been done there. Also, the United States became a homeland for many of those who were able to survive. Secondly, however, we must share the responsibility for not being willing to

acknowledge forty years ago that this horrible event was occurring. Finally, because we are humane people, concerned with the human rights of all peoples, we feel compelled to study the systematic destruction of the Jews so that we may seek to learn how to prevent such enormities from occurring in the future.[31]

Not only would the memory mandated by this museum include the lives of "new Americans," but it would reinforce America's self-idealization as a haven for the world's oppressed. It would serve as a universal warning against the bigotry and antidemocratic forces underpinning such a catastrophe and call attention to the potential in all other totalitarian systems for such slaughter.

As a national landmark, the U.S. Holocaust Memorial Museum would necessarily plot the Holocaust according to the nation's own ideals, its pluralist tenets. In the words of the memorial council, therefore, the Holocaust began "before a shot was fired, with persecution of Jews, dissenters, blacks, Gypsies, and the handicapped. The Holocaust gathered force as the Nazis excluded groups of people from the human family, denying them freedom to work, to study, to travel, to practice a religion, claim a theory, or teach a value. This Museum will illustrate that the loss of life itself was but the last stage in the loss of all rights."[32] In being defined as the ultimate violation of America's Bill of Rights and as the persecution of plural groups, the Holocaust encompasses all the reasons immigrants—past, present, and future—ever had for seeking refuge in America.

When cultural critics protested that such a museum, though necessary, would be a blight on the mall, the U.S. Holocaust Memorial Council countered, "This Museum belongs at the center of American life because as a democratic civilization America is the enemy of racism and its ultimate express, genocide. An event of universal significance, the Holocaust has special importance for Americans: in act and word the Nazis denied the deepest tenets of the American people." That is, the U.S. Holocaust Memorial defines what it means to be American by graphically illustrating what it means not to be American. As a reminder of "the furies beyond our shores," in one columnist's words, the museum would define American existence in the great distance between "here" and "there."[33]

This would be the beginning of what the museum's Project Director Michael Berenbaum termed the "Americanization of the Holocaust." In Berenbaum's words, the museum's story of the Holocaust would have to be

> told in such a way that it would resonate not only with the survivor in New York and his children in Houston or San Francisco, but with a black leader from Atlanta, a mid-western farmer, or a northeastern industrialist. Millions of Americans make

pilgrimages to Washington; the Holocaust Museum must take them back in time, transport them to another continent, and inform their current reality. The Americanization of the Holocaust is an honorable task provided that the story told is faithful to the historical event.[34]

Of course, as Berenbaum also makes clear, the story itself depends entirely on who is telling it—and to whom.

As it turns out, by mandating a national Holocaust memorial museum, President Carter not only Americanized the Holocaust, but he also set a new national standard for suffering. After seeing the Holocaust formally monumentalized on the mall, visitors may begin to view it less as an actual historical event and more as an ideal of catastrophe against which all other past and future destructions might be measured—or pitted. Moreover, the museum has issued an implicit challenge to two other long-suffering American ethnic groups, African and Native Americans, who have responded by proposing their own national institutions for the mall and nearby. Indeed, when informed that the National African American Museum would be located in an existing building of the Smithsonian Institute, Illinois Representative Gus Savage responded angrily that since "Jews and Indians had their own place on the Mall," so should African Americans.[35] Given the mall's own dark past as the former site of holding pens for newly arrived African slaves, the National African American Museum will be placed in an especially difficult position: Not only will it be asked to share an authentic site of African American suffering with other groups, but it will also be faced with the unenviable task of teaching Americans that the topographical center of their national shrine is also the site of America's greatest, ineradicable shame.

Conclusion

Rather than continuing to insist that memorial forms like the monument do what modern societies, by dint of their vastly heterogeneous populations and competing memorial agendas, will not permit them to do, I've long believed that the best way to save the monument, if it's worth saving at all, is to enlarge its life and texture to include its genesis in historical time, the activity that brings a monument into being, the debates surrounding its origins, its production, its reception, its life in the mind. That is to say, rather than seeing polemics as a by-product of a nation's memorial mandate, I would make the polemics surrounding these mandates' existence their central animating feature. For I believe that in our age of heteroglossia

(Bakhtin's term), national mandates for the memory of catastrophe succeed only insofar as they allow themselves full expression of the debates, arguments, and tensions generated in the noisy give-and-take among competing constituencies driving their very creation. In this view, memory as represented in these mandates might also be regarded as a never to be completed process, animated (not disabled) by the forces of history bringing it into being.

In this spirit, I believe the mandate for memory now being issued by the Lower Manhattan Development Corporation in its memorial guidelines for a World Trade Center site memorial includes the public discussion and dissent that continue to drive the process forward. That is, they are attempting to build into this site a worldview that allows for competing, even conflicting, agendas—and make this, too, part of the memorial process. Rather than fretting about the appearance of disunity (all memorial processes are exercises in disunity, even as they strive to unify memory), we should make our questions and the public debate itself part of our memory work. Memory is, after all, a process and is everlasting only when it remains a process and not a finished result. For just as the memory of the September 11 attacks on the World Trade Center is a negotiation between past and present, it is also an ongoing negotiation among all the groups of people whose lives were affected by this event and those whose lives will be shaped by what is built there. How will the needs of competing constituencies, all with profoundly legitimate concerns, be balanced against each other? If, as part of a well-defined process, the debates are conducted openly and publicly, they will be as edifying as they are painful.

The firefighters and their families; the police officers and their families; the office workers, traders, and bankers from around the world; city officials; developers; architects; artists; immediate neighbors and outlying neighbors; and tourists may all want something a little different in this site. And the hundreds of victims from ninety-two other nations, all with their own reasons for being there, will be remembered according to their own national memorial mandates. There must be room in both the process and in the actual commemorative site for the memory of all these people and their families. Memory at ground-zero is not zero-sum: It is an accumulation of all these disparate experiences and needs. Just as we accommodate ourselves to the competing needs of others every day of our lives in New York City, just as we live together but separately, come and go together but separately, we must build into both the process and this site the capacity to remember together but separately.

In conclusion, I find that there may never be an intrinsic link between a memorial mandate and its formal expression. Such links are created, not born, I believe, and it is in the exchange between a mandate and the expressions giving it form that finally constitute the memorial itself. In the case of the World Trade Center Memorial, a memorial mandate was issued by an advisory council composed of family members of victims, downtown residents, and firefighters and police officers associations. But their mandate was not intended to prescribe a particular kind of memory for the masses; rather, it was explicitly invoked as a mandate for the memorial itself. The "Memorial Mission Statement" thus reads:

> Remember and honor the thousands of innocent men, women, and children murdered by terrorists in the horrific attacks of February 26, 1993 and September 11, 2001. Respect this place made sacred through tragic loss. Recognize the endurance of those who survived, the courage of those who risked their lives to save others, and the compassion of all who supported us in our darkest hours. May the lives remembered, the deeds recognized, and the spirit reawakened be eternal beacons, which reaffirm respect for life, strengthen our resolve to preserve freedom, and inspire an end to hatred, ignorance and intolerance.[36]

Once again, how will this mandate actually be expressed in material forms? Can all of these things ever be shown, or must they always be told? On balance, I believe these things are expressed by the juxtaposition of mandate and a day, of the pairing of a mandate and the forms we find to give it expression. Finally, I find that in the exchange between mandate and its concretized form, we force each to accommodate the other and thereby force memory to evolve over time, to remain alive in the tension between what each can and cannot be accomplished by either one. Just as I've hoped to animate memorials by restoring the noisy debates and discussions that brought them into being, we might now animate memory itself by granting that even when mandated, it demands expression finally in our own responses to it, that our lives as lived in response to these national memorial mandates may be the best formal expression of memory itself.

Notes

1. See Maurice Halbwachs, *Les Cadres sociaux de la memoire* (Paris: Presses Universitaires de France, 1952).

2. This section on Holocaust Remembrance Day and a subsequent section on the U.S. Holocaust Memorial Museum adapt and elaborate parts of my *The Texture of Memory: Holocaust Memorials and Meaning* (New Haven, CT: Yale University Press, 1993).

3. For elaboration of this distinction, see Gerard Genette, *Narrative Discourse: An Essay in Method* (Ithaca, NY: Cornell University Press, 1980), 33–35.

4. For keen insight into Israel's "counter-tradition," see Yael Zerubavel, "Invented Tradition and Counter-tradition: The Social Construction of the Past in Israeli Culture," paper presented at the Association for Jewish Studies annual meeting, December 1990. Also see Zerubavel's "New Beginning, Old Past: The Collective Memory of Pioneering in Israeli Culture," in *New Perspectives on Israeli History: The Early Years of the State*, Laurence J. Silberstein, ed. (New York: New York University Press, 1990), 193–215; and "The Holiday Cycle and the Commemoration of the Past: History, Folklore and Education," in *The Proceedings of the Ninth World Congress of Jewish Studies*, vol. 2 (1986), 111–18.

5. The Tenth of Teveth commemorates the beginning of the siege; the 17th of Tammuz marks the first breach in the walls of the city; the Ninth of Av recalls the destruction of the Temple; and the Third of Tishri remembers the assassination of Gedaliah, the governor of Judah appointed by Nebuchadnezzar. For details surrounding the historical origins of these dates, see Theodor H. Gaster, *Festivals of the Jewish Year: A Modern Interpretation and Guide* (New York: Morrow, Quill, 1978), 194–96.

6. In a further example, it is possible that the Chmielnicky massacres commenced on precisely the same day as the Blois blood libel murder of thirty-two Jews 477 years earlier, as is traditionally believed. But it is more likely that when the anniversary of the Blois massacre in 1171 became a day of fasting for Jewish communities in England, France, and the Rhineland, it also became the anniversary for subsequent massacres occurring in the same general period on the calendar.

7. From Yad, Hilchot Ta'aniyot 5:1.

8. See Irving Greenberg, *The Jewish Way* (New York: Summit Books, 1988), 330. Contrary to Greenberg's suggestion that this date never caught on with religious Jews, the Tenth of Teveth is, in fact, widely observed by much of the ultra-Orthodox community in Israel, where jahrzeit candles are kindled and ceremonies are conducted at the Mount of Olives cemetery.

9. For a much more extended discussion of the statists' attitude toward the Holocaust, see Charles S. Liebman and Eliezer Don-Yehiya, *Civil Religion in Israel* (Berkeley: University of California Press, 1983), 107.

10. *Divrei HaKnesset* 1951, 1657.

11. For a much more detailed anthropological analysis of the calendar's narrative, see Don Handelman, *Models and Mirrors: Towards an Anthropology of Public Events* (Cambridge: Cambridge University Press, 1990), 194–200. In his rich study, Handelman also reminds us that in falling seven days before Yom Hazikkaron, Yom Hashoah recapitulates the Jewish mourning period (shiva) of seven days.

12. Liebman and Don-Yehiya, *Civil Religion in Israel*, 100–118.

13. *Divrei HaKnesset* 1959, 1386.

14. See Saul Friedlander, "Die Shoah als Element in der Konstruktion Israelischer Erinnerung," *Babylon* 2 (1987), 10–22; "The Shoah Between Memory and History," *Jerusalem*

Quarterly 53 (Winter 1990); and "Roundtable Discussion," in *Writing and the Holocaust,* Berel Lang, ed. (New York: Holmes and Meier, 1988), 288.

15. Irving Greenberg also recalls that in 1984, Rabbi Pinchas Teitz proposed yet another alternative date for Yom Hashoah: the anniversary of Hitler's death—that is, the 17th of Iyar, the day before Lag B'omer, the festival celebrating the end of the Sfirah period; *The Jewish Way,* 332–33.

16. Quoted from "Day of Memorial for Victims of the European Jewish Disaster and Heroism—27 Nissan, 5719," *Yad Vashem Bulletin 4/5* (October 1959), 27.

17. As quoted in Liebman and Don-Yehiya, *Civil Religion in Israel,* 178 (emphasis added).

18. As quoted in Liebman and Don-Yehiya, *Civil Religion in Israel,* 184.

19. We might recall in this context that the Yad Vashem World Council's first convention was held on April 19, 1957 (anniversary of the Warsaw Ghetto Uprising), which fell that year on the 8th of Iyar and not on the 27th of Nissan, at Har Hazikkaron. After reading a number of letters, Chairman Benzion Dinur asked all to rise for a moment's silence. According to one report, "the council rose in a minute's silence *in memory of the victims of the European holocaust and of those who fell in the defense of the homeland.*" In this equation, martyrs and fighters are united here by the memory of those who were both. See "Yad Washem World Council Convenes on Memorial Hill in Jerusalem," *Yad Vashem Bulletin* no. 1 (April 1957), 31 (emphasis added).

20. See Shmuel Spector, "Yad Vashem," *Encyclopedia of the Holocaust,* vol. 4 (New York: Macmillan, 1990), 1681–86.

21. For the Hebrew transcript of part of the debate, see *Divrei HaKnesset* (Minutes of the Parliament) 1953, 131–54.

22. From Nachum Goldman, "The Influence of the Holocaust on the Change in the Attitude of World Jewry to Zionism and the State of Israel," in *Holocaust and Rebirth: A Symposium* (Jerusalem: Yad Vashem, 1974), 103.

23. From "Martyrs' and Heroes' Remembrance (Yad Vashem) Law, 5713–1953," reprinted fully in *Yad Vashem: The Holocaust Martyrs' and Heroes' Remembrance Authority, Jerusalem* (Jerusalem: Yad Vashem, 1986), 4. This law is also translated and reprinted in the *State of Israel Yearbook* (Jerusalem: Government Printing Press, 1954), 250–51.

24. See "Department for the Registration of the Martyred," *Yad [V]ashem Bulletin* (April 1957), 40.

25. Idit Zertal, *From Catastrophe to Power: Holocaust Survivors and the Emergence of Israel* (Berkeley: University of California Press, 1998).

26. From the Jewish Telegraphic Agency press bulletin, December 2, 1942. I am grateful to Lucia Ruedenberg for alerting me to this reference in her "'Remember 6,000,000': Civic Commemoration of the Holocaust in New York City," unpublished doctoral dissertation, 1994.

27. See Atay Citron, "Pageantry and Theatre in the Service of Jewish Nationalism in the United States: 1933–1946," unpublished doctoral dissertation, New York University, 1989.

28. Quoted in "City Rejects Park Memorials to Slain Jews," *The New York Times*, February 11, 1965, 1; and in "2 Jewish Monuments Barred from Park," *New York World Telegram and Sun*, February 10, 1965, 1.

29. For more on the political dimension of this memorial museum, see Michael Berenbaum, *After Tragedy and Triumph: Modern Jewish Thought and the American Experience* (Cambridge: Cambridge University Press, 1991), 3–16. For an explicit record of the museum's conceptualization and development, see Edward Linenthal, *Preserving Memory: The Struggle to Create America's Holocaust Museum* (New York: Viking/Penguin, 1995).

30. Charles Maier, *The Unmasterable Past: History, Holocaust, and German National Identity* (Cambridge, MA: Harvard University Press, 1988), 165.

31. From an undated press release of the U.S. Holocaust Memorial Council.

32. *The Campaign for the United States Holocaust Memorial Museum*, published by the U.S. Holocaust Memorial Council, 4.

33. George Will, "Holocaust Museum: Antidote for Innocence," *The Washington Post*, March 10, 1983.

34. Michael Berenbaum, *After Tragedy and Triumph*, 20.

35. Quoted in Cassandra Burrell, "Supporters of African-American Museum Object to Smithsonian Control," Associated Press, September 15, 1992.

36. Statement courtesy of the Lower Manhattan Development Corporation as found in its Memorial Design Competition brochure and Web site, http://www.renewnyc.com.

Index

The Amherst Series in Law, Jurisprudence, and Social Thought

EDITED BY
Austin Sarat, Lawrence Douglas, and Martha Merrill Umphrey

Law and Catastrophe (2007)
Law and the Sacred (2007)
How Law Knows (2007)
The Limits of Law (2005)
Law on the Screen (2005)